JIMD Reports
Volume 7

Johannes Zschocke · K. Michael Gibson
Editors-in-Chief

Garry Brown · Eva Morava
Editors

Verena Peters
Managing Editor

JIMD Reports –
Case and Research Reports, 2012/4

Editor-in-Chief
Johannes Zschocke
Medizinische Universität Innsbruck
Sektionen für Humangenetik und Klinische
Innsbruck
Austria

Editor-in-Chief
K. Michael Gibson
Michigan Technological University
Biological Sciences
Houghton Michigan
USA

Editor
Garry Brown
University of Oxford
Department of Biochemistry
Genetics Unit
Oxford
United Kingdom

Editor
Eva Morava
Radboud University Nijmegen
Medical Center
Department of Pediatrics
IGMD
Nijmegen
Netherlands

Managing Editor
Verena Peters
Center for Child and Adolescent
Medicine
Heidelberg University Hospital
Heidelberg
Germany

ISSN 2192-8304   ISSN 2192-8312 (electronic)
ISBN 978-3-642-32441-3   ISBN 978-3-642-32442-0 (eBook)
DOI 10.1007/978-3-642-32442-0
Springer Heidelberg New York Dordrecht London

Printed on acid-free paper

Springer is part of Springer Science+Business Media (www.springer.com)

Contents

JIMD Reports
DOI 10.1007/8904_2012_128

Necrotizing Enterocolitis and Respiratory Distress Syndrome as First Clinical Presentation of Mitochondrial Trifunctional Protein Deficiency

Eugène F. Diekman · Carolien C.A. Boelen · Berthil H.C.M.T. Prinsen · Lodewijk IJlst · Marinus Duran · Tom J. de Koning · Hans R. Waterham · Ronald J.A. Wanders · Frits A. Wijburg · Gepke Visser

Received: 4 August 2011 / Revised: 12 January 2012 / Accepted: 13 January 2012 / Published online: 31 March 2012
© SSIEM and Springer-Verlag Berlin Heidelberg 2012

Abstract *Background:* Newborn screening (NBS) for long-chain 3-hydroxy acyl-CoA dehydrogenase (LCHAD) deficiency does not discriminate between isolated LCHAD deficiency, isolated long-chain keto acyl-CoA (LCKAT) deficiency and general mitochondrial trifunctional protein (MTP) deficiency. Therefore, screening for LCHAD deficiency inevitably comprises screening for MTP deficiency, which is much less amenable to treatment. Furthermore, absence of a clear classification system for these disorders is still lacking.

Communicated by: Piero Rinaldo

Competing interests: None declared

Electronic supplementary material: The online version of this chapter (doi:10.1007/8904_2012_128) contains supplementary material, which is available to authorized users.

E.F. Diekman · L. IJlst · M. Duran · H.R. Waterham · R.J.A. Wanders
Laboratory Genetic Metabolic Diseases, Department of Clinical Chemistry, Academic Medical Center, University of Amsterdam, Amsterdam, The Netherlands

E.F. Diekman · B.H.C.M.T. Prinsen · T.J. de Koning · G. Visser (✉)
Department of Metabolic Diseases and Endocrine Diseases, Wilhelmina Children's Hospital, KC 03.063.0, University Medical Centre Utrecht, Lundlaan 6, 3584 EA, Utrecht, The Netherlands
e-mail: gvisser4@umcutrecht.nl

C.C.A. Boelen
LUMC Leiden University Medical Center, Leiden, The Netherlands

R.J.A. Wanders · F.A. Wijburg
Department of Pediatrics, Emma Children's Hospital, Academic Medical Center, University of Amsterdam, Amsterdam, The Netherlands

Materials and Methods: Two newborns screened positive for LCHAD deficiency died at the age of 10 and 31 days, respectively. One due to severe necrotizing enterocolitis (NEC), cardiomyopathy and multiorgan failure and the other due to severe infant respiratory distress syndrome (IRDS) and hypertrophic cardiomyopathy. (Keto)-acylcarnitine concentration and enzymatic analysis of LCHAD and LCKAT suggested MTP deficiency in both patients. Mutation analysis revealed a homozygous HADHB c.357+5delG mutation in one patient and a homozygous splice-site HADHB mutation c.212+1G>C in the other patient.

Data on enzymatic and mutation analysis of 40 patients with presumed LCHAD, LCKAT or MTP deficiency were used to design a classification to distinguish between these disorders.

Discussion: NEC as presenting symptom in MTP deficiency has not been reported previously. High expression of long-chain fatty acid oxidation enzymes reported in lungs and gut of human foetuses suggests that the severe NEC and IRDS observed in our patients are related to the enzymatic deficiency in these organs during crucial stages of development.

Furthermore, as illustrated by the cases we propose a classification system to discriminate LCHAD, LCKAT and MTP deficiency based on enzymatic analysis.

Introduction

In 2007, the newborn screening (NBS) program in the Netherlands was expanded with 13 inborn errors of metabolism, including the autosomal recessive long-chain fatty acid oxidation (FAO) disorder and long-chain

3-hydroxy acyl-CoA dehydrogenase (LCHAD) deficiency. LCHAD is part of mitochondrial trifunctional protein (MTP), which harbours two additional enzymes in long-chain FAO: long-chain enoyl-CoA hydratase (LCEH) and long-chain keto acyl-CoA thiolase (LCKAT). The enzyme active sites are located on different subunits, named alpha- and beta-, which together form an octameric complex of 4α- and 4β-subunits (Kamijo et al. 1994; Ushikubo et al. 1996). LCHAD and LCEH are located on the α-subunit, and are both encoded by the *HADHA* gene. LCKAT is located on the β-subunit, and is encoded by the *HADHB* gene (Kamijo et al. 1994; Ushikubo et al. 1996).

NBS for LCHAD deficiency is performed by measuring C16-OH-carnitine levels in dried blood spots. However, discrimination between isolated general MTP deficiency, LCHAD deficiency, isolated LCKAT deficiency (LCKAT deficiency) or isolated LCEH deficiency (LCEH deficiency; not identified yet) cannot be made on the basis of the acylcarnitine profile, but requires specific enzyme testing in lymphocytes or fibroblasts. Furthermore measurement of 3-keto-C18:1-carnitine and 3-keto-C18:2-carnitine, which accumulate in case of LCKAT deficiency but not LCHAD deficiency, might also be helpful (Wanders et al. 2010).

By far the most common mutation associated with LCHAD deficiency is the HADHA c.1528G>C mutation (p.Glu510Gln, allelic frequency 60%) (Das et al. 2006). Mutations associated with a deficient activity of all enzymes (MTP deficiency) are more heterogeneous.

While no patients with isolated LCEH deficiency have been reported, MTP deficiency has been reported in relatively large series of patients (Choi et al. 2007; den Boer et al. 2003; Olpin et al. 2005; Purevsuren et al. 2009; Saudubray et al. 1999; Spiekerkoetter et al. 2003; Ushikubo et al. 1996). Patients with MTP deficiency, including isolated LCHAD deficiency, most often present with acute metabolic decompensation consisting of hypoketotic hypoglycemia and rhabdomyolysis, generally followed by cardiomyopathy and later peripheral neuropathy. Hypotonia, areflexia and hepatic encephalopathy have also been described (Choi et al. 2007; den Boer et al. 2003; Olpin et al. 2005; Purevsuren et al. 2009; Saudubray et al. 1999; Spiekerkoetter et al. 2003; Ushikubo et al. 1996). In contrast, isolated LCKAT deficiency appears extremely rare and only one patient, who presented with lethal cardiorespiratory failure, has been reported (Das et al. 2006).

We present two patients identified by NBS with an abnormal screening result suggestive for LCHAD deficiency, who were subsequently diagnosed with MTP deficiency. Both patients were already severely ill at the time the results of the NBS became available, and showed remarkable clinical symptoms which are generally not observed in patients with a defect in FAO.

It is not possible to distinguish isolated LCHAD deficiency and LCKAT deficiency of MTP deficiency based on NBS results, clinical signs and symptoms or DNA mutation analysis. Therefore, we propose a classification based on enzymatic analysis of LCHAD and LCKAT.

Materials and Methods

Case 1

The patient, a girl, was the first child of consanguineous Caucasian parents. The pregnancy was complicated by eclampsia. The mother had 7 seizures, ALAT of 29 (normal 0–35), ASAT of 36 (0–30), and low platelets 127×10^{-9}/L (normal 150–450). The pregnancy was therefore terminated at 35 weeks of gestation by caesarean section. Birth weight was 2,110 g (−0.4SD), length 40 cm (<−2.5SD) and head circumference 30 cm (<−2.5SD). APGAR scores were 4, 8 and 8 after 1, 5 and 10 min, respectively. Postpartum glucose was 9.6 mmol/L (normal 3.6–5.6), lactate 7.1 mmol/L (normal 0.0–2.2), ammonia 127 μmol/L (normal 0–75), LDH 899 U/L (normal 0–250), ASAT/ALAT were normal. CK at day 7 was 478 U/L (normal 0–145). With normal intake, plasma glucose levels remained above 4.2 mmol/L and lactate levels decreased to 2.6 mmol/L. On day 3, she had rectal blood loss. Upon suspicion of a necrotizing enterocolitis (NEC), she was admitted to the neonatal intensive care unit and parenteral feeding was initiated. On day 7, a sudden clinical deterioration suggested a gut perforation as a complication of the NEC and a laparotomy was performed. No perforation was found, but intestinal biopsies later showed the classical pathology of a NEC. There was no clinical improvement observed and echocardiography revealed a severely dilated cardiomyopathy with low cardiac output. On this same day (day 7), the NBS results from a dried blood spot, taken at day 4, became available and showed an elevated C16-OH-carnitine suggestive of LCHAD deficiency. Additional analysis in plasma revealed increased concentrations of hydroxy-acylcarnitines (Table 1). Subsequently, keto-acylcarnitine concentrations were analyzed and showed increased 3-keto-C18:1-carnitine and 3-keto-C18:2-carnitine, which is suggestive of LCKAT deficiency (Table 1). Enzymatic analysis showed reduced activities of both LCHAD and LCKAT (lymphocytes). DNA mutation analysis of the *HADHB* gene (GenBank accession number BC066963) showed a homozygous splice-site mutation in intron 4, c.212+1G>C, predicted to lead to aberrant splicing of the HADHB mRNA transcript. No mutation was found in the HADHA gene.

The patient developed seizures during prolonged hypotensive episodes. Cerebral ultrasound studies showed

Table 1 Acylcarnitine profile (NBS and plasma), enzymatic activity and mutations of both patients (controls ± SD)

Acylcarnitine profile (bloodspot)	Patient 1: 35 weeks, 2,110 g (µmol/L) (day 4)	Patient 2: 30 weeks, 1,275 g (µmol/L) (day 3)	Control values (µmol/L)
C16-OH-carnitine	1.44	0.63	≤0.08
Acylcarnitine profile (plasma)	Patient 1 (µmol/L) (day 7)	Patient 2 (µmol/L) (day 7)	Control values (µmol/L) ($N = 700$)
Free carnitine	6.9	9.82	22.35–54.80
C14-carnitine	0.2	0.36	0–0.08
C14:1-carnitine	0.1	0.42	0.02–0.18
C14:1-OH-carnitine	0.05	0.11	0–0.02
C16-carnitine	1.13	1.33	0.06–0.24
C16-OH-carnitine	0.59	0.44	0–0
C16:1-carnitine	0.32	0.45	0.02–0.08
C16:1-OH-carnitine	0.19	0.25	0–0.02
C18:1-OH-carnitine	0.6	1.12	0–0.02
3-Keto-C18:1-carnitine	Detected	Detected	Undetectable
3-Keto-C18:2-carnitine	Detected	Detected	Undetectable
Activity		Patient 1 (nmol/(min.mg protein))	Patient 2 (nmol/(min.mg protein))	Control values (nmol/(min.mg protein))
HAD activity	C16	12[L] (23%)	10[L] (19%)	53 ± 18[L] ($N = 88$)
			11[F] (14%)	79 ± 16[F] ($N = 215$)
	C4	103[L] (69%)	116[L] (78%)	149 ± 46[L] ($N = 135$)
			86[F] (76%)	113 ± 29[F] ($N = 215$)
	C16/ C4	0.12[L]	0.08[L]	0.37 ± 0.20[L] ($N = 88$)
			0.13[F]	0.72 ± 0.14[F] ($N = 215$)
LCKAT activity		0.9[L] (8%)	1.7[F] (9%)	10.2 ± 3.6[L] ($N = 41$)
				18.3 ± 5.4[F] ($N = 215$)
Mutation analysis		Patient 1	Patient 2	
HADHA gene (allele 1)		Normal	Normal	
HADHA gene (allele 2)		Normal	Normal	
HADHB gene (allele 1)[a]		c.212+1G>C	c.357+5delG	
HADHB gene (allele 2)[a]		c.212+1G>C	c.357+5delG	

L lymphocytes, *F* fibroblasts

[a] Numbering according to GenBank sequence BC066963

minimal flaring and a minor bleeding (grade II). Despite vigorous treatment, including high dose (8–10 mg/kg/min) intravenous glucose infusion, she died at 10 days of age because of severe multiorgan failure.

Case 2

The patient, a boy, was the first child of consanguineous parents. Pregnancy was complicated by pre-eclampsia and intrauterine growth retardation (IUGR). The pregnancy was terminated at 30 weeks by caesarean section because of foetal distress. Birth weight was 1,275 g (<−1.0SD), length 37 cm (−2.0SD) and head circumference 28 cm (1-0SD). APGAR scores were 7, 8 and 9 after 1, 5 and 10 min, respectively. On day 1, he became hypotensive and developed severe infant respiratory distress syndrome (IRDS; grades III and IV) despite multiple surfactant administrations. He was artificially ventilated. Despite continuous glucose infusion (8–10 mg/kg/min), he had multiple hypoglycaemic episodes (postpartum glucose 1.8 mmol/L) and a persistent lactic acidosis (>10 mmol/L, normal 0.0–2.2). Echocardiography performed on day 6 revealed a hypertrophic cardiomyopathy. Cerebral ultrasound showed no abnormalities. On day 10, the NBS results from a dried blood spot, taken at day 3, became available which revealed an elevated C16-OH-carnitine, suspicious for LCHAD deficiency. Additional analysis in plasma revealed increased concentrations of hydroxy-acylcarnitines (Table 1). Subsequently, keto-acylcarnitine concentrations were analyzed and showed increased 3-keto-C18:1-carnitine and 3-keto-C18:2-carnitine, which is suggestive of LCKAT deficiency (Table 1). Enzymatic analysis showed reduced activities of both LCHAD and LCKAT (fibroblasts). DNA mutation analysis of the *HADHB* gene (GenBank accession number BC066963) showed a homozygous mutation c.357+5delG, which was subsequently shown by cDNA analysis to result in a complete skipping of exon 6. No mutation was found in the HADHA gene.

Despite extensive treatment including ventilatory support, glucose infusion, parenteral feeding, carnitine supplementation (100 mg/kg/day), long-chain triglyceride restriction and medium-chain enriched feeding, he died 31 days after birth because of respiratory failure.

Classification

To be able to discriminate between LCHAD, LCKAT and MTP deficiency, we retrospectively analyzed data of 40 nonrelated patients in whom LCHAD and LCKAT activity was measured and DNA mutation analysis was performed. In addition, we analyzed data of 215 subjects in whom LCHAD and LCKAT activity was measured because of suspected FAO disorder, but in whom no FAO defect was found.

Hydroxyacyl-CoA dehydrogenase (HAD) activity was measured in homogenates of cultured fibroblasts by observing the decrease in absorbance at 340 nm (Wanders et al. 1990). 3-Keto-hexadecanoyl-CoA (C16) and acetoacetyl-CoA

(C4) have been used as substrates for LCHAD activity measurements. LCHAD shows highest activity with C16 as substrate with virtually no reactivity with C4 as substrate (Wanders 1990; Wanders et al. 2010). However, another dehydrogenase – short chain hydroxyaxcyl-CoA dehydrogenase (SCHAD) – has activity with both C16 and C4. To be able to acquire an accurate approximation of LCHAD activity, the activities with C16 and C4 as substrate were used as a ratio (C16/C4).

A full deficiency of LCHAD will result in a C16/C4 HAD activity ratio of approximately 0.2, which is characteristic of the SCHAD enzyme, as it is five times more active with C4 than C16 as a substrate. A C16/C4 HAD activity ratio higher than approximately 0.2 is a result of (residual) LCHAD activity.

LCKAT activity was measured in homogenates of cultured fibroblasts by following the decrease in absorbance at 303 nm (Wanders et al. 1990).

All enzyme analyses were performed on a Cobas Centrifugal Analyzer (Hofmann-La Roche, Basel, Switzerland).

DNA was extracted from blood leukocytes and amplified by polymerase chain reaction (PCR). After amplification, 20 exons of HADHA and 16 exons of HADHB gene were sequenced.

The results are shown in Fig. 1.

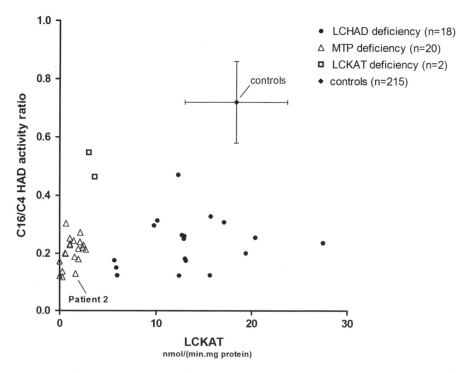

Fig. 1 C16/C4 HAD activity ratio and LCKAT activity of patients and controls. C16/C4 HAD activity ratio and LCKAT activity have been analyzed in fibroblasts of 40 patients with HADHA and/or HADHB mutations and 215 controls as described in "Materials and Methods" section. Patients can be divided into three groups: LCHAD (*filled circle*), LCKAT (*open square*) and MTP (*open triangle*) deficiency. Mean C16/C4 HAD activity ratio of control group 0.72 ± 0.14 (mean ± SD; *vertical error bars*). Mean LCKAT activity of control group 18.3 ± 5.4 nmol/(min.mg protein) (mean ± SD; *horizontal error bars*)

Analytical Methods

Enzymatic activity measurements in lymphocytes and/or fibroblast were carried out as previously described (Wanders et al. 1990/1992). NBS acylcarnitine profiling was performed as described by Chace et al. (2003). Plasma acylcarnitine profiling was performed as described by Minkler et al. (2005). In addition, keto-acylcarnitine profiling was performed by incubating 50 μL of plasma with 5 μL of MOX Reagent (Pierce, Rockford, IL, USA) containing 2% methoxyamine·HCl in pyridine. The mixture was left to stand at room temperature for 2 h to allow the formation of methoxime-derivatives of all keto-containing substances. Following this incubation, the samples were treated in the usual manner for the isolation and derivatization of acylcarnitines, which were analyzed as their butyl-ester derivatives (Minkler et al. 2005).

Discussion

NEC as a presenting symptom in MTP deficiency has, to our knowledge, not been reported previously. Severe IRDS has been recognized as a rare early symptom in MTP deficiency (Olpin et al. 2005). Cardiomyopathy, which was present in both patients, is a frequently reported complication in MTP deficiency (den Boer et al. 2003; Olpin et al. 2005; Purevsuren et al. 2009; Spiekerkoetter et al. 2003).

Although it is generally assumed that FAO plays no or only a minor role during intrauterine life due to the abundance of glucose provided by the mother via the placenta (Oey et al. 2006; Shekhawat et al. 2003), the clinical course in our patients, as well as the patient with isolated LCKAT deficiency described by Das et al. (2006), suggests that normal function of the MTP complex is needed for normal intestinal and pulmonary development and function. Berger and Wood showed that complete disruption of long-chain FAO at the level of LCAD in animal models results in increased embryonic mortality (Berger and Wood 2004). It has also been shown that FAO enzymes are expressed abundantly in human placentas (Shekhawat et al. 2003). Furthermore, patients with long-chain defects in FAO may already display cardiomyopathy before and immediately after birth (den Boer et al. 2003; Olpin et al. 2005; Purevsuren et al. 2009; Spiekerkoetter et al. 2004), demonstrating a role for long-chain FAO during intrauterine life. Finally, the (pre)eclampsia, the premature delivery and the foetal distress seen in both hereby described patients, is also in line with this hypothesis (Oey et al. 2005).

While NEC is a relatively common complication in ill premature babies, it is rarely seen in newborns of 35 weeks gestational age, with a birth weight >1,500 g and in absence of a history of hypovolemic shock and/or asphyxia (Neu and Walker 2011). In addition, severe IRDS, not responding to multiple administration of surfactant, is rare in neonates born after 30 weeks gestation. We therefore hypothesize that both the NEC observed in patient 1 and the severe IRDS in patient 2 are linked to the defective long-chain FAO. Early foetal expression patterns of long-chain FAO enzymes, including VLCAD and LCHAD, demonstrate that these enzymes are expressed not only in myocardial tissue, but also abundantly in the foetal lung and gut (Oey et al. 2005). MTP deficiency during intrauterine life may therefore interfere with normal development or maturation of the foetal intestine and lungs. In the gut, this might result in decreased mucus synthesis, decreased intracellular junction integrity and increased permeability, both potentially related to the development of NEC (Neu and Walker 2011). In addition, intestinal villous atrophy and inflammation is observed in carnitine transport-deficient (OCTN2) mice (Shekhawat et al. 2007).

In the lungs, surfactant is secreted by alveolar type II cells, and decreased maturation or functioning of this process may lead to IRDS (Olpin et al. 2005). We therefore believe that MTP deficiency during intrauterine life may hamper normal surfactant synthesis.

We identified two novel mutations in the *HADHB* gene. The mutation found in patient 1, c.212+1G>C, affects the splice-donor site of intron 4 and is predicted to result in skipping of exon 4. Because fibroblasts were not available from patient 1, this could not be studied at the cDNA level. The G deletion at position c.357+5 found in patient 2 results in skipping of exon 6, as demonstrated by cDNA analysis prepared from mRNA isolated from fibroblasts of the patient.

Both mutations did not only result in a markedly reduced LCKAT activity, but also affected enzyme activity of LCHAD, which is possibly due to the loss of integrity of the MTP complex (Ushikubo et al. 1996). It is known that a single mutation in the *HADHA* or *HADHB* gene can result in either an isolated deficiency of LCHAD or LCKAT, or reduced activity of both enzymes. However, until now it has not been possible to clearly distinguish isolated LCHAD and LCKAT deficiency from MTP deficiency. We propose a classification system based on the C16/C4 HAD activity ratio and LCKAT residual enzyme activities measured in 40 patients. As shown in Fig. 1, the patients can be divided into three groups, which we have labelled as LCHAD, LCKAT and MTP deficiency. The LCHAD group contains patients with a mean C16/C4 HAD activity ratio of 0.24 ± 0.09 (mean $\pm$ SD) combined with a keto-thiolase activity of 13.5 ± 5.4 nmol/(min.mg protein) (mean $\pm$ SD). The MTP deficiency group consists of patients with a mean C16/C4 HAD activity ratio of 0.2 ± 0.05 (mean $\pm$ SD) and an LCKAT activity of 1.3 ± 0.88 nmol/(min.mg protein) (mean $\pm$ SD). The third group, isolated LCKAT deficiency, includes two patients with a C16/C4 HAD activity ratio of approximately 0.5 and an LCKAT activity of <5 nmol/(min.mg protein) (Fig. 1). The activity of the different enzymes could not be fully predicted based on the mutations. Although the LCHAD group includes 17 of the 18 patients homozygous for the HADHA c.1528G>C mutation, the MTP-deficient group includes patients with either alpha- or beta-subunit mutations. The two isolated LCKAT-deficient patients, one of which was reported previously (Das et al. 2006), have distinct beta-subunit mutations. Based on this classification, patient 2 will be categorized as MTP deficient. We could not obtain fibroblasts of the first patient and are therefore unable to classify this patient unambiguously. However, predicted is that the MTP protein of patient 1 is absent, because the mutation resulted in exon skipping. We therefore conclude patient 1 is also MTP deficient.

In summary, we present two patients in whom NBS results were consistent with LCHAD deficiency. They were eventually diagnosed with MTP deficiency, based on a novel proposed classification system. One of the patients presented with a severe NEC, which has not been associated previously with long-chain FAO defects. Furthermore, a severe IRDS was observed in the other patient. Both clinical presentations may be explained by high expression patterns of long-chain FAO enzymes not only in myocardial tissue, but in lung and gut tissue as well. Deficiency of MTP in the gut and lung might therefore explain the development of severe NEC and IRDS in addition to the cardiomyopathy found in our patients.

Acknowledgments Eugene Diekman is paid by a grant of ZonMW, The Netherlands Organisation for Health Research and Development, dossier 200320006.

Synopsis

MTP-deficient patients may present with NEC or IRDS. Illustrated by two cases with these symptoms, a classification system for LCHAD, LCKAT and MTP deficiency based on enzymatic analysis is proposed.

Conflict of Interest

None.

References

Berger PS, Wood PA (2004) Disrupted blastocoele formation reveals a critical developmental role for long-chain acyl-CoA dehydrogenase. Mol Genet Metab 82:266–272

Chace DH, Kalas TA, Naylor EW (2003) Use of tandem mass spectrometry for multianalyte screening of dried blood specimens from newborns. Clin Chem 49:1797–1817

Choi J-H, Yoon H-R, Kim G-H et al (2007) Identification of novel mutations of the HADHA and HADHB genes in patients with mitochondrial trifunctional protein deficiency. Int J Mol Med 19:81–87

Das AM, Illsinger S, Lücke T et al (2006) Isolated mitochondrial long-chain ketoacyl-CoA thiolase deficiency resulting from mutations in the HADHB gene. Clin Chem 52:530–534

Den Boer MEJ, Dionisi-Vici C, Chakrapani A et al (2003) Mitochondrial trifunctional protein deficiency: a severe fatty acid oxidation disorder with cardiac and neurologic involvement. J Pediatr 142:684–689

Kamijo T, Wanders RJ, Saudubray JM et al (1994) Mitochondrial trifunctional protein deficiency. Catalytic heterogeneity of the mutant enzyme in two patients. J Clin Invest 93:1740–7

Minkler PE, Ingalls ST, Hoppel CL (2005) Strategy for the isolation, derivatization, chromatographic separation, and detection of carnitine and acylcarnitines. Anal Chem 77:1448–57

Neu J, Walker WA (2011) Necrotizing enterocolitis. N Eng J Med 364:255–264

Oey NA, Den Boer MEJ, Wijburg FA et al (2005) Long-chain fatty acid oxidation during early human development. Pediatr Res 57:755–759

Oey NA, Ruiter JPN, Attié-Bitach T et al (2006) Fatty acid oxidation in the human fetus: implications for fetal and adult disease. J Inherit Metab Dis 29:71–75

Olpin SE, Clark S, Andresen BS et al (2005) Biochemical, clinical and molecular findings in LCHAD and general mitochondrial trifunctional protein deficiency. J Inherit Metab Dis 28:533–544

Purevsuren J, Fukao T, Hasegawa Y et al (2009) Clinical and molecular aspects of Japanese patients with mitochondrial trifunctional protein deficiency. Mol Genet Metab 98:372–377

Saudubray JM, Martin D, De Lonlay P et al (1999) Recognition and management of fatty acid oxidation defects: a series of 107 patients. J Inherit Metab Dis 22:488–502

Shekhawat P, Bennett MJ, Sadovsky Y et al (2003) Human placenta metabolizes fatty acids: implications for fetal fatty acid oxidation disorders and maternal liver diseases. Am J Physiol Endocrinol Metab 284:1098–105

Shekhawat PS, Srinivas SR, Matern D et al (2007) Spontaneous development of intestinal and colonic atrophy and inflammation in the carnitine-deficient jvs (OCTN2(−/−)) mice. Mol Genet Metab 92:315–24

Spiekerkoetter U, Sun B, Khuchua Z et al (2003) Molecular and phenotypic heterogeneity in mitochondrial trifunctional protein deficiency due to beta-subunit mutations. Hum Mut 21:598–607

Spiekerkoetter U, Khuchua Z, Yue Z et al (2004) General mitochondrial trifunctional protein (TFP) deficiency as a result of either alpha- or beta-subunit mutations exhibits similar phenotypes because mutations in either subunit alter TFP complex expression and subunit turnover. Pediatr Res 55:190–196

Ushikubo S, Aoyama T, Kamijo T et al (1996) Molecular characterization of mitochondrial trifunctional protein deficiency: formation of the enzyme complex is important for stabilization of both alpha- and beta-subunits. Am J Hum Genet 58:979–988

Wanders RJ, Ijlst L, van Gennip AH et al (1990) Long-chain 3-hydroxyacyl-CoA dehydrogenase deficiency: identification of a new inborn error of mitochondrial fatty acid beta-oxidation. J Inherit Metab Dis 13:311–4

Wanders RJ, Ijlst L, Poggi F et al (1992) Human trifunctional protein deficiency: a new disorder of mitochondrial fatty acid beta-oxidation. Biochem Biophys Res Commun 188:1139–1145

Wanders RJA, Ruiter JPN, Ijlst L et al (2010) The enzymology of mitochondrial fatty acid beta-oxidation and its application to follow-up analysis of positive neonatal screening results. J Inherit Metab Dis 33:479–494

JIMD Reports
DOI 10.1007/8904_2012_132

CASE REPORT

Temporal Intradiploic Dilative Vasculopathy: An Additional Pathogenic Factor for the Hearing Loss in Fabry Disease?

Carla Pinto Moura · Carlos Soares · Daniela Seixas ·
Margarida Ayres-Bastos · João Paulo Oliveira

Received: 19 November 2011 / Revised: 22 January 2012 / Accepted: 07 February 2012 / Published online: 24 March 2012
© SSIEM and Springer-Verlag Berlin Heidelberg 2012

Abstract Fabry disease (FD) is caused by progressive accumulation of neutral glycosphingolipids, including in ganglion neural and vascular endothelial cells, as a result of lysosomal α-galactosidase deficiency. High frequencies progressive sensorineural hearing loss (HL), sudden deafness, tinnitus and dizziness are otological symptoms frequently reported.

A 45-year-old man with FD, on haemodialysis since age 25, complaining of progressive HL, was started on enzyme replacement therapy (ERT) because of cardiac complications. A bilateral sloping sensorineural HL was found at baseline audiological evaluation. Computed tomography of the ears showed enlargement of the intradiploic vascular

Communicated by: Verena Peters

Competing interests: None declared

C.P. Moura
Department of Otolaryngology, Faculty of Medicine, University
of Porto and Hospital São João, Alameda Hernâni Monteiro,
4200-319 Porto, Portugal
e-mail: cmoura@med.up.pt

C.P. Moura · J.P. Oliveira (✉)
Department of Genetics, Faculty of Medicine, University of Porto
and Hospital São João, Alameda Hernâni Monteiro,
4200-319 Porto, Portugal
e-mail: jpo@med.up.pt

C. Soares
Hemodialysis Unit - Riba de Ave, DIAVERUM, Rua Padre
Narciso Melo 14, 4765–259 Riba de Ave, Portugal
e-mail: csoares@portugalmail.com

D. Seixas · M. Ayres-Bastos
Department of Neuro-Radiology, Faculty of Medicine, University
of Porto and Hospital São João, Alameda Hernâni Monteiro,
4200-319 Porto, Portugal
e-mail: dseixas@med.up.pt; mail.mab@clix.pt

channels, principally in the petrous bone. The magnetic resonance angiography showed elongation and ectasia of the middle cerebral arteries and the arteries of the Circle of Willis, particularly the internal carotid and the basilar arteries. Follow-up audiological evaluations documented progressive worsening of HL, mainly in the high frequencies range, despite high dose ERT and evidence of cardiac improvement.

The intradiploic vascular abnormalities of the temporal bones reported herein have never been described in association with FD and may have contributed to the pathogenesis of progressive HL, by a 'stealing' effect upon the cochlear blood supply (like in cavernous haemangioma of the internal auditory meatus), in addition to the other mechanisms of ischaemic injury to the Organ of Corti described in FD. This clinical observation shows the value of comprehensive neuroimaging investigation of HL in FD and emphasizes the importance of early institution of specific therapy, before the occurrence of irreversible inner ear lesions and hearing damage.

Introduction

Fabry disease (FD) is a rare X-linked metabolic disorder due to accumulation of globotriaosylceramide (GL-3) and other neutral glycosphingolipids (GSL) in many organs and tissues, as a result of deficient activity of the lysosomal enzyme α-galactosidase (Desnick et al. 2001, 2003). Involvement of the microvascular endothelium with progressive luminal narrowing and occlusion leading to tissue ischaemia is regarded as a major pathogenic mechanism of FD (Desnick et al. 2001, 2003).

Life-threatening complications of FD in males include chronic kidney disease (CKD), hypertrophic cardiomyopathy

⌂ Springer

and cerebrovascular disease (Desnick et al. 2001, 2003). Although the age of onset of clinical symptoms and the disease severity may vary widely, CKD invariably progresses to end-stage renal disease (ESRD) in men with the classical phenotype of FD. Heterozygous females show variable organ involvement (Wilcox et al. 2008) but seldom are as severely affected as males (Desnick et al. 2001, 2003).

Otological symptoms of FD include progressive sensorineural hearing loss (HL), sudden deafness, tinnitus, dizziness and vertigo (Germain et al. 2002; Conti and Sergi 2003; Hajioff et al. 2003a, b; Hegemann et al. 2006; Ries et al. 2007; Palla et al. 2007; Keilmann et al. 2009; Sergi et al. 2010). The prevalence of HL and tinnitus associated with FD increases with age, males being usually affected earlier and more severely than females. On pure tone audiometry (PTA), up to 88% of adult males with FD may show evidence of progressive HL (Palla et al. 2007). The HL is more frequent and severe at the high-tone frequencies (Hajioff et al. 2003a, b; Palla et al. 2007). High-frequency HL may occur as an isolated PTA finding in children reporting subjective hearing impairment (Keilmann et al. 2009), as well as in young adults with clinically normal audition (Germain et al. 2002). Residual α-galactosidase activity lowers the risk of HL by about twofold in comparison with patients who have no detectable enzyme activity (Ries et al. 2007). A decreased number of spiral ganglia, atrophy of the stria vascularis and of the spiral ligament in all cochlear turns, loss of hair-cells mainly of the basal turns and GSL accumulation in vascular endothelial and ganglion cells of the ear were the histopathological features described in the temporal bones of two middle-aged FD patients with sensorineural HL (Schachern et al. 1989). These findings can explain most of the otological symptoms associated with FD.

Two genetically engineered human α-galactosidase preparations (agalsidase alfa and agalsidase beta) are available for the treatment of FD by enzyme replacement therapy (ERT) (Eng et al. 2001; Schiffmann et al. 2001). Despite minor glycosylation differences (Lee et al. 2003), the two agalsidases showed similar biodistribution in the mouse model of FD (Lee et al. 2003) and apparently have identical immunogenicity in humans (Linthorst et al. 2004; Vedder et al. 2007). Overall results of ERT for up to 60 months suggest that patients with mild to moderate degrees of HL may slightly improve with ERT, whereas hearing does not change significantly in patients with normal hearing or in those with severe HL, as evaluated by PTA at baseline (Hajioff et al. 2003a, b; Hegemann et al. 2006; Palla et al. 2007).

Herein, we report the significant worsening of the hearing thresholds of a classically affected male with FD while on ERT with agalsidase beta for more than 3 years and the radiological finding of vascular abnormalities in the temporal bones that have never been described in association with FD and may have contributed to the HL in this patient.

Case Report

A 46-year-old male with ESRD and classic FD, hemizygous for the α-galactosidase missense mutation p.C94S, with no detectable residual enzyme activity on the leukocyte and plasma assays, was referred for comprehensive otolaryngological assessment before the beginning of ERT with agalsidase beta. Although the patient complained of bilateral tinnitus and slow progressive impairment in word discrimination for several years, he had refused proper otological assessment 2 months before, when he reported sudden subjective deterioration of hearing. He was dialysis dependent since the age of 25 years. Persistent sinus bradycardia with haemodynamic intolerance to ultrafiltration during the dialysis sessions had been noted for several months. Despite his blood pressure being well controlled with no need for anti-hypertensive medication, the echocardiography showed a mildly dilated left atrium (diameter = 44 mm) and mild left ventricular hypertrophy, with a maximum left ventricular wall thickness of 12 mm at the interventricular septum. There was no past history of transient ischaemic attacks (TIA) or stroke events.

A detailed otolaryngological medical history was obtained at baseline, along with a thorough physical and instrumental examination of the ears, nose and throat (ENT). The patient reported no symptoms of ear fullness, dizziness or instability, and denied previous history of cranial trauma, otological surgery, noise exposure or use of ototoxic drugs. There was no family history of deafness. The only abnormality found on the ENT physical examination was a bilateral inferior turbinates hypertrophy. The baseline audiogram showed bilateral sloping sensorineural HL, most severe at the 4–8 kHz range (Fig. 1). Distortion product and transient-evoked otoacoustic emissions were absent. The tympanogram was normal bilaterally, with absent acoustic reflexes. The patient refused hearing aid.

Treatment with agalsidase beta was started with 35 mg (~0.8 mg/kg) infused every other week during a haemodialysis session (Kosch et al. 2004). Eleven months later, the unit dose was eventually doubled to 70 mg (~1.6 mg/kg), for lack of cardiac improvement and reversion of plasma GL-3 to pre-ERT levels. The increase in plasma GL-3 levels was coincidental with the first detection of anti-agalsidase IgG antibodies. The evolution of the plasma GL-3 levels, the anti-agalsidase beta antibody titre and the most relevant echocardiographic parameters, up to 30 months after the institution of ERT, are shown in

Audiometric Evolution

Fig. 1 The baseline audiogram showed bilateral sloping sensorineural HL, with a pure tone threshold average for 0.5 kHz, 1 kHz and 2 kHz of 43.3 dB bilaterally and pure tone thresholds of 60 and 75 dB at 4 kHz and of 75 and 85 dB at 8 kHz, respectively on the right and the left ear. In the follow-up audiogram, 40 months later, pure tone threshold averages for 0.5, 1 and 2 kHz were 65 dB in the right ear and 56.7 dB in the left ear; thresholds at 4 and 8 kHz were both 85 dB in the right ear and 100 and 90 dB, respectively, in the left ear. Audiometry of higher frequencies (8 to 20 kHz) showed a HL of 85 dB at the 8 to 12.5 kHz tone range in the right ear and of 90 dB in the left ear, with slightly better results at the 16 and 20 kHz range, bilaterally

Table 1 Evolution of plasma GL-3 levels, anti-agalsidase beta antibodies titer and echocardiographic parameters while on ERT with agalsidase beta

Time on enzyme replacement therapy

	Baseline	1 month	2 months	3 months	6 months	9 months	12 months	24 months	30 months
Plasma GL-3[a]	6.6	6.3	4.3	4.4	6.4	–	6.5	5.7	–
IgG antibodies[b]	–	–	–	–	(+)	–	1/400	1/200	–
Echocardiographic parameters									
LVIDD	48	–	–	–	–	50	–	54	58
PWTD	10	–	–	–	–	11	–	11	10
IVSTD	12	–	–	–	–	13	–	8	7
LVMI	69	–	–	–	–	83	–	68	67

Blood samples for GL-3 assay in plasma and for anti-agalsidase antibodies detection in serum were always collected at the start of the corresponding hemodialysis session

[a] GL-3 plasma levels, normal range: 1.6–3.3 µmol/l (Laboratory Medicine / Clinical Chemistry, Sahlgren University Hospital, Molndal, Sweden)

[b] Anti-agalsidase beta IgG antibodies (Genzyme Corporation, Cambridge, MA 02142, USA)

Echocardiographic dimensions (mm): *LVIDD* Left Ventricular Internal Diameter in Diastole, *PWTD* Posterior Wall Thickness in Diastole, *IVSTD* Interventricular Septum Thickness in Diastole, *LVMI* Left Ventricular Mass Index, expressed as g/m^2.7

Table 1. During this period, the patient did not miss any of the scheduled agalsidase beta infusions.

A follow-up audiogram, performed 42 months after the start of ERT, including testing for the 8–20 kHz pure tone frequency range, documented a significant worsening of the hearing impairment (Fig. 1). In addition, comprehensive imaging data of the brain, ears, temporal bones and adjacent skull structures were obtained by computerized tomography (CT) and magnetic resonance (MR) with MR angiography (MRA). Enlargement of intradiploic vascular channels of the temporal bones, particularly in the mastoids, petrous pyramids and adjacent to the temporomandibular joints, was noted on the CT scan of the base of the skull (Fig. 2a, b). Similar lesions were present in the frontal and sphenoid bones. The middle and inner ear on both sides had no abnormalities. There were bilateral atheromatous calcifications of the vertebral, basilar (BA) and internal carotid arteries (ICAs). The MRA (Fig. 3) showed elongation and ectasia of the middle cerebral arteries and the arteries of the Circle of Willis, particularly the ICAs and the BA. The ICAs displayed luminal irregularities. On T2-weighted brain MR imaging scan (Fig. 4) there were multiple focal areas of high signal

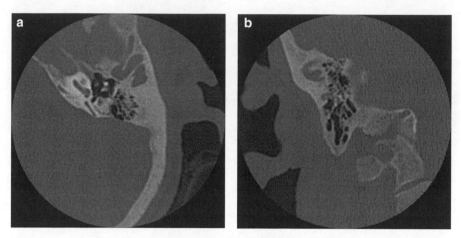

Fig. 2 (**a, b**) The CT scan of the base of the skull showed enlargement of intradiploic vascular channels distributed bilaterally in the temporal bones, particularly in the mastoids, petrous pyramids, and adjacent to the temporomandibular joints

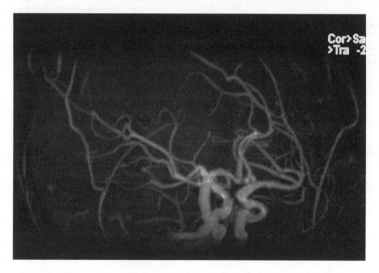

Fig. 3 The MRA showed elongation and ectasia of the middle cerebral arteries and the arteries of the Circle of Willis, particularly the internal carotid arteries and the basilar artery. The internal carotid arteries displayed luminal irregularities

intensity in the periventricular white matter and corona radiata bilaterally as well as in the right thalamus. At this time, the patient accepted hearing aid.

Discussion

Progressive sensorineural HL and sudden deafness, occurring unilaterally or bilaterally, are frequently observed in men with the classic form of FD (Germain et al. 2002; Conti and Sergi 2003; Hajioff et al. 2003a, b; Hegemann et al. 2006; Ries et al. 2007; Palla et al. 2007; Keilmann et al. 2009; Sergi et al. 2010). Although the otological symptoms and the HL are not life-threatening complications of FD, they may have a profound impact on the patients' quality of life

(Mehta 2003). The lack of histopathological evidence of GSL accumulation in the spiral ganglia (Schachern et al. 1989) and the frequent occurrence of sudden deafness are suggestive that the HL in FD is more probably due to ischaemic injury to the organ of Corti rather than to involvement of the neural hearing pathways.

The HL in FD patients correlates with the decline of renal function, as well as with clinical or brain imaging evidence of cerebrovascular disease (Germain et al. 2002). In our patient, the explanation for the HL might be confounded by the long-term haemodialysis treatment, as sensorineural HL, especially in the high frequencies range, is quite prevalent among patients on renal replacement therapies (Quick 1976; Kligerman et al. 1981; Stavroulaki et al. 2001). However, the HL associated with advanced

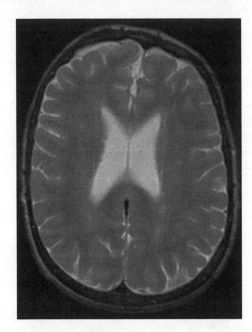

Fig. 4 The T2-weighted brain MR imaging scan there were multiple focal areas of high signal intensity in the periventricular white matter and corona radiata bilaterally as well as in the right thalamus

CKD is not clinically relevant in most cases and does not appear to deteriorate with time in long-term haemodialysis patients (Bazzi et al. 1995). Furthermore, a careful review of the patient's medical records excluded past exposure to the ototoxic drugs that commonly contribute to the risk of HL in the dialysis setting (Quick 1976).

Large-vessel ectasia and large-vessel and small-vessel occlusive disease, predominantly affecting the posterior circulation, are typical brain imaging and angiographic findings in FD (Crutchfield et al. 1998; Fellgiebel et al. 2006), even in patients with no previous history of TIA or stroke. Dolichoectasia of cerebral arteries is a recognized risk factor for ischaemic stroke and small-vessel disease in the general population (Ince et al. 1998; Pico et al. 2005). Blood flow within the dolichoectatic arteries is turbulent and has a retrograde stream component that may compromise the irrigation of distal tissues. Therefore, as the main arterial blood supply of the human cochlea is through the internal auditory artery, which is a branch from the BA, the organ of Corti stands in a high-risk vascular region for ischaemic injury in FD. Our patient had no clinical history of cerebrovascular events but his brain CT, MR and MRA findings were characteristic of those observed in middle-aged classically affected men with FD. Kidney dysfunction is independently related to asymptomatic increase in ICA and BA diameters as evaluated by brain MR imaging in elderly pre-dialysis CKD patients (Ichikawa et al. 2009), but this is of uncertain clinical significance in ESRD patients. Furthermore, we are not aware of any studies showing a clinically significant risk of intracranial dilative arteriopathy in ESRD (with the exception of cerebral aneurysms in patients with Autosomal Dominant Polycystic Kidney Disease), or of case reports of intradiploic dilative arteriopathy in long-term haemodialysis patients.

To the best of our knowledge, enlargement of intra-diploic vessels has never been reported in FD patients. The vascular lacunae affecting the temporal, frontal and sphenoid bones observed in our patient resemble those described in intraosseous haemangiomas of the cranium (Bastug et al. 1995). Although it cannot be conclusively demonstrated that the vascular abnormalities identified in our patient are a complication of FD, their multifocality and association with dolichoectasia of the ICAs and BA, and the rarity of cranial intraosseous haemangiomas, suggest that they might be an expression of the dilative arteriopathy of FD. Angiomas of the petrous bone are a cause of rapidly progressive, severe sensorineural HL (Dufour et al. 1994). Stealing of cochlear blood supply by the angiomatous lesion has been suggested as a possible mechanism for ischaemic damage to the organ of Corti in HL associated with a cavernous haemangioma of the internal auditory meatus (Madden and Sirimina 1990). By analogy, we postulate that the vascular malformations of the temporal bones might have contributed to the hearing impairment in our patient. It is also possible that the dilative arteriopathy of the temporal bones is more frequent than so far recognized in FD patients presenting with severe HL, because its diagnostic assessment is not usually done as thoroughly as we did in our case. Furthermore, as the angiectatic bone lesions were observed more than 2 years after treatment with high dose agalsidase beta, which proved effective in reducing the left ventricular mass, most probably they do not regress with ERT.

References

Bastug D, Ortiz O, Schoclet S (1995) Hemangiomas in the calvaria: imaging findings. AJR Am J Roentgenol 164:683–687

Bazzi C, Venturini CT, Pagani C, Arrigo G, D'Amico G (1995) Hearing loss in short-and long-term haemodialysed patients. Nephrol Dial Transplant 10:1865–1868

Conti G, Sergi B (2003) Auditory and vestibular findings in Fabry disease: a study of hemizygous males and heterozygous females. Acta Pediatr Suppl 92(443):33–37, discussion 27

Crutchfield KE, Patronas NJ, Dambrosia JM et al (1998) Quantitative analysis of cerebral vasculopathy in patients with Fabry disease. Neurology 50:1746–1749

Desnick RJ, Ioannou YA, Eng CM (2001) Alpha-galactosidase A deficiency: Fabry disease. In: Scriver CR, Beaudet AL, Sly WS, Valle D, Kinzler KE, Vogelstein B (eds) The metabolic and molecular bases of inherited disease. McGraw Hill, New York, pp 3733–3774

Desnick RJ, Brady R, Barranger J et al (2003) Fabry disease, an under-recognized multisystemic disorder: expert recommendations for diagnosis, management, and enzyme replacement therapy. Ann Intern Med 138:338–436

Dufour JJ, Michaud LA, Mohr G, Pouliot D, Picard C (1994) Intratemporal vascular malformations (angiomas): particular clinical features. J Otolaryngol 23:250–253

Eng CM, Guffon N, Wilcox WR, International Collaborative Fabry Disease Study Group et al (2001) Safety and efficacy of recombinant human alpha-galactosidase A-replacement therapy in Fabry's disease. N Engl J Med 345:9–16

Fellgiebel A, Müller MJ, Ginsberg L (2006) CNS manifestations of Fabry's disease. Lancet Neurol 5:791–795

Germain D, Avan P, Chassaing A, Bonfils P (2002) Patients affected with Fabry disease have an increased incidence of progressive hearing loss and sudden deafness: an investigation of twenty-two hemizygous male patients. BMC Med Genet 3:10

Hajioff D, Goodwin S, Quiney R, Zuckerman J, MacDermot KD, Mehta A (2003a) Hearing improvement in patients with Fabry disease treated with agalsidase alfa. Acta Paediatr Suppl 92 (443):28–30, discussion 27

Hajioff D, Enever Y, Quiney R, Zuckerman J, MacDermot K, Mehta A (2003b) Hearing loss in Fabry disease: the effect of agalsidase alfa replacement therapy. J Inher Metab Dis 26:787–794

Hegemann S, Hajioff D, Conti G et al (2006) Hearing loss in Fabry disease: data from the Fabry Outcome Survey. Eur J Clin Invest 36:654–662

Ichikawa H, Takahashi N, Mukai M, Katoh H, Akizawa T, Kawamura M (2009) Intracranial dilative arteriopathy is associated with chronic kidney disease and small vessel diseases in the elderly. J Stroke Cerebrovasc Dis 18:435–442

Ince B, Petty GW, Brown RD Jr, Chu CP, Sicks JD, Whisnant JP (1998) Dolichoectasia of the intracranial arteries in patients with first ischemic stroke: a population-based study. Neurology 50:1694–1998

Keilmann A, Hajioff D, Ramaswami U, FOS Investigators (2009) Ear symptoms in children with Fabry disease: data from the Fabry Outcome Survey. J Inherit Metab Dis 32:739–744

Kligerman AB, Solangi KB, Ventry IM, Goodman AI, Weseley SA (1981) Hearing impairment associated with chronic renal failure. Laryngoscope 91:583–592

Kosch M, Koch HG, Oliveira JP et al (2004) Enzyme replacement therapy administered during hemodialysis in patients with Fabry disease. Kidney Int 66:1279–1282

Lee K, Jin X, Zhang K et al (2003) A biochemical and pharmacological comparison of enzyme replacement therapies for the glycolipid storage disorder Fabry disease. Glycobiology 13:305–313

Linthorst G, Hollak C, Donker-Koopman W, Strijland A, Aerts J (2004) Enzyme therapy for Fabry disease: neutralizing antibodies toward agalsidase alpha and beta. Kidney Int 66:1589–1595

Madden GJ, Sirimina KS (1990) Cavernous hemangioma of the internal auditory meatus. J Otolaryngol 19:288–291

Mehta A (2003) Commentary. Acta Paediatr Suppl 92(443):27

Palla A, Hegemann S, Widmer U, Straumann D (2007) Vestibular and auditory deficits in Fabry disease and their response to enzyme replacement therapy. J Neurol 254:1433–1442

Pico F, Labreuche J, Touboul PJ, Leys D, Amarenco P (2005) Intracranial arterial dolichoectasia and small-vessel disease in stroke patients. Ann Neurol 57:472–479

Quick C (1976) Hearing loss in patients with dialysis and renal transplants. Ann Otol 85:776–790

Ries M, Kim HJ, Zalewski CK et al (2007) Neuropathic and cerebrovascular correlates of hearing loss in Fabry disease. Brain 130:143–150

Schachern P, Shea D, Paparella M, Yoon T (1989) Otologic histopathology of Fabry's disease. Ann Otol Rhinol Laryngol 98:359–363

Schiffmann R, Kopp JB, Austin HA et al (2001) Enzyme replacement therapy in Fabry disease: a randomized controlled trial. JAMA 285:2743–2749

Sergi B, Conti G, Paludetti G, Interdisciplinary Study Group On Fabry Disease (2010) Inner ear involvement in Anderson-Fabry disease: long-term follow-up during enzyme replacement therapy. Acta Otorhinolaryngol Ital 30:87–93

Stavroulaki P, Nikolopoulos TP, Psarommatis I, Apostolopoulos N (2001) Hearing evaluation with distortion-product otoacoustic emissions in young patients undergoing haemodialysis. Clin Otolaryngology 26:235–242

Vedder AC, Linthorst GE, Houge G et al (2007) Treatment of Fabry disease: outcome of a comparative trial with agalsidase alfa or beta at a dose of 0.2 mg/kg. PLoS One 2:e598

Wilcox WR, Oliveira JP, Hopkin RJ et al (2008) Females with Fabry disease frequently have major organ involvement: lessons from the Fabry Registry. Mol Genet Metab 93:112–128

JIMD Reports
DOI 10.1007/8904_2012_133

Hereditary Intrinsic Factor Deficiency in Chaldeans

Amy C. Sturm · Elizabeth C. Baack · Michael B. Armstrong · Deborah Schiff ·
Ayesha Zia · Sureyya Savasan · Albert de la Chapelle · Stephan M. Tanner

Received: 20 December 2011 / Revised: 06 February 2012 / Accepted: 08 February 2012 / Published online: 18 March 2012
© SSIEM and Springer-Verlag Berlin Heidelberg 2012

Abstract Juvenile vitamin B_{12} or cobalamin (Cbl) deficiency is notoriously difficult to explain due to numerous
acquired and inherited causes. The consequences of
insufficient Cbl are megaloblastic anemia, nutrient malabsorption, and neurological problems. The treatment is
straightforward with parenteral Cbl supplementation that
resolves most health issues without an urgent need to
clarify their cause. Aside from being clinically unsatisfying,
failing to elucidate the basis of Cbl deficiency means
important information regarding recurrence risk is not
available to the individual if the cause is contagious or
inherited. Acquired causes have largely disappeared in the
Modern World because they were mostly due to parasites or
malnutrition. Today, perhaps the most common causes of
juvenile Cbl deficiency are Imerslund-Gräsbeck syndrome
and inherited intrinsic factor deficiency (IFD). Three genes
are involved and genetic testing is complicated and not
widely available. We used self-identified ancestry to
accelerate and confirm the genetic diagnosis of IFD in
three families of Chaldean origin. A founder mutation
limited to Chaldeans from Iraq in the intrinsic factor gene
GIF was identified as the cause. World events reshape the
genetic structure of populations and inherited diseases in
many ways. In this case, all the patients were diagnosed in
the USA among recent immigrants from a single region.
While IFD itself is not restricted to one kind of people,
certain mutations are limited in their range but migrations
relocate them along with their host population. As a result,
self-identified ancestry as a stratifying characteristic should
perhaps be considered in diagnostic strategies for rare
genetic disorders.

Communicated by: Matthias Baumgartner

Competing interests: None declared

A.C. Sturm · A. de la Chapelle · S.M. Tanner (✉)
Human Cancer Genetics Program, Comprehensive Cancer Center,
The Ohio State University, BRT 804, 460W. 12th Ave,
Columbus, OH 43210, USA
e-mail: stephan.tanner@osumc.edu

A.C. Sturm · E.C. Baack
Human Genetics, Department of Internal Medicine, The Ohio
State University, Columbus, OH 43210, USA

M.B. Armstrong
Division of Hematology-Oncology, Department of Pediatrics,
Duke University, Durham, NC 27710, USA

D. Schiff
Division of Hematology/Oncology, Rady Children's Hospital,
UCSD School of Medicine, San Diego, CA 92123, USA

A. Zia · S. Savasan
Division of Hematology/Oncology, Children's Hospital of
Michigan, Wayne State University School of Medicine, Detroit,
MI 48201, USA

Introduction

Juvenile cobalamin deficiency (JCD) can be attributed to
dietary, infectious, or hereditary causes (Gräsbeck 2006).
In modern societies, dietary reasons and infectious agents
are largely a thing of the past. As a result, inborn causes of
JCD such as cobalamin (Cbl) malabsorption, serum
transport defects (transcobalamin deficiency), or metabolic
mutations have become increasingly important in pediatric
care (Watkins and Rosenblatt 2011). Cbl deficiency leads
to various hematological problems that range from mild
weakness to life-threatening acute megaloblastic anemia.
Neurological symptoms are often overlooked since they
can vary from mild learning difficulties to overt antisocial
outbursts and may be recognized only if carefully sought

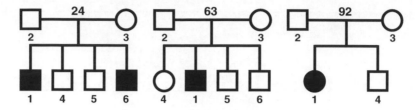

Fig. 1 Pedigrees of four patients affected by inherited intrinsic factor deficiency (*IFD*, *filled symbols*) due to the intronic mutation c.1073+5G>A in the intrinsic factor gene, *GIF*. The parents are all of Chaldean ancestry

for (Gräsbeck 2006; Luder et al. 2008). Due to the myriad of potential causes and varying degree of symptoms, explaining the etiology of JCD can be daunting and diagnostic testing can take months (Carmel et al. 2003; Watkins and Rosenblatt 2011). Even if malabsorption of Cbl is established as the cause of JCD by measurement of transcobalamin-bound Cbl (Obeid and Herrmann 2007) or with the Schilling test (Schilling 1953), clarification remains challenging and clinical testing for gastric causes can be invasive. Genetically, pathogenic mutations in either *CUBN* (Aminoff et al. 1999) or *AMN* (Tanner et al. 2003) cause Imerslund-Gräsbeck syndrome (IGS; OMIM261100; Imerslund 1960; Gräsbeck et al. 1960) and in intrinsic factor deficiency (IFD; OMIM261000; Katz et al. 1972), mutations in the *GIF* gene are causative (Yassin et al. 2004; Tanner et al. 2005). Proteinuria is sometimes indicative of IGS but lack thereof is not necessarily diagnostic for IFD (Gräsbeck and Tanner 2011), while mono-symptomatic proteinuria due to *CUBN* mutations is not necessarily associated with Cbl deficiency (Ovunc et al. 2011). Although all three genes can be sequenced (Tanner et al. 2004, 2005; Storm et al. 2011), the screening task is complex given the genetic heterogeneity.

Recent human migrations have an impact on the clinical and epidemiological picture of diseases that are encountered in the Western World. While infectious diseases come to mind first, hereditary conditions similarly move about with migrants. Resulting population bottlenecks in resettled communities might lead to founder effects that change the incidence and the mutational spectrum of disorders in the new home country. The current focus on personalized medicine has to adapt to that challenge. Self-identified ethnicity might offer rapid shortcuts for diagnostics in certain cases if the ancestry of a patient is carefully evaluated. We present the results of genetic testing for JCD in three families of Chaldean origin. While the first family with two patients had to wait several years for the analysis to be completed because the significance of the mutation was not obvious, genetic confirmation in a second family permitted the last patient to be diagnosed within weeks. Thus, patients with rare disorders might be readily and reliably diagnosed if their ethnicity is taken into account.

Patients and Methods

Patients and Samples

Family 24 came to our attention because of two boys (ages 14 and 8) with macrocytic anemia (mean cell volume 111.6 fL, normal range 86-98 fL) and Cbl deficiency (exact values were not reported, Fig. 1 and Table 1). No neurological problems were noted and nutritional deficiency was excluded. The macrocytosis and anemia were responsive to parenteral Cbl replacement therapy. A Schilling test in the older sibling (24-1) was positive for malabsorption but the younger boy (24-6) was not tested. The older brother did not respond to the addition of intrinsic factor (IF) during the second stage of the Schilling test (Schilling 1953). Proteinuria was absent in both brothers. The family reported Chaldean ancestry and lives in the USA. Family history was significant for a paternal cousin reportedly with a similar problem, although he also had neurologic symptoms that were not further specified.

Family 63: A 4-year-old boy, one of four siblings, was noted to have Cbl deficiency (Cbl 74 pg/mL) (Fig. 1). Intramuscular injections with Cbl resulted in a significant improvement from his anemia and macrocytosis. He was then lost to follow-up until 3 years later when he presented after an episode of syncope. Hemoglobin at that time was 11.1 gm/dL, mean cell volume (MCV) was elevated at 103 fL, and the Cbl level was severely decreased at <70 pg/mL. A screen for Anti-IF antibodies was negative. Proteinuria was absent and nutritional deficiency was excluded. Family history was significant for multiple cousins who also received regular Cbl injections. He was restarted on monthly parenteral Cbl with hydroxocobalamin 1,000 mcg. On this regimen, his hemoglobin has remained normal, and his MCV and Cbl level have normalized. Both parents of family 63 were born in the same region of Northern Iraq and reported no consanguinity. The family resides in the USA and they identify themselves as Chaldeans.

Family 92: A 4-year-old Iraqi Chaldean girl was seen with a diagnosis of acute lymphoblastic leukemia (ALL) in Michigan, USA (manuscript in preparation). She was originally diagnosed in Turkey and had received chemotherapy there until recently. She had a history of Cbl

Table 1 Genetic analysis of *GIF* in three Chaldean families with inherited intrinsic factor deficiency

Sample ID	Family information	*GIF* intron 7[a]	*GIF* M1	*GIF* M3b	*GIF* M4	*GIF* M5	*GIF* M6	*GIF* M7
24-1	Male patient	hom c.1073+5G>A	176-176	235-235	160-160	256-256	204-204	238-238
24-2	Father	het c.1073+5G>A	176-176	235-239	160-160	nd	nd	nd
24-3	Mother	het c.1073+5G>A	176-176	235-235	160-160	nd	nd	nd
24-4	Brother	het c.1073+5G>A	176-176	235-235	160-160	nd	nd	nd
24-5	Brother	c.1073+5G	176-176	235-239	160-160	nd	nd	nd
24-6	Brother, patient	hom c.1073+5G>A	176-176	235-235	160-160	256-256	204-204	238-238
63-1	Male patient	hom c.1073+5G>A	176-176	235-235	160-160	256-256	204-204	238-238
63-2	Father	het c.1073+5G>A	176-176	223-235	160-160	nd	nd	nd
63-3	Mother	het c.1073+5G>A	176-176	235-243	160-160	nd	nd	nd
63-4	Sister	het c.1073+5G>A	176-176	235-243	160-160	nd	nd	nd
63-5	Brother	c.1073+5G	176-176	223-243	160-160	nd	nd	nd
63-6	Brother	het c.1073+5G>A	176-176	235-243	160-160	nd	nd	nd
92-1	Female patient	hom c.1073+5G>A	176-176	235-235	160-160	256-256	204-204	238-238
92-2	Father	het c.1073+5G>A	nd	nd	nd	nd	nd	nd
92-3	Mother	het c.1073+5G>A	nd	nd	nd	nd	nd	nd
92-4	Brother	c.1073+5G	nd	nd	nd	nd	nd	nd

hom homozygous, *het* heterozygous, *nd* not done

[a] Numbering relative to adenine in the ATG start codon of *GIF* (GenBank accession AP002347)

deficiency, which was diagnosed in Iraq when she was 18 months of age and was treated there for 4–5 months by injections. Until the diagnosis of ALL in February 2011, she had not received any further treatment for Cbl deficiency, though she was given an oral Cbl supplement monthly during chemotherapy before relocating to the USA according to the parents. Repeated Cbl measurements were low at 161 and 92 pg/mL (normal low = 211) and she had elevated serum and urine methylmalonic acid levels (0.74 mmol/L (normal <0.4) and 9.6 mmol/mol creatinine (normal <3.6), respectively). She had no proteinuria as the urine microalbumin/creatinine ratio was within normal limits. Her diet was appropriate.

Methods

Peripheral blood samples were collected after informed consent was obtained under Ohio State University review board protocol 2005H0201 according to the Declaration of Helsinki. Genomic DNA samples were isolated from peripheral blood by standard phenol-chloroform-ethanol-precipitation. For mutational analysis, families 24 and 63 were studied for suspected IGS (*AMN* and *CUBN*) and IFD (*GIF*) as previously described (Tanner et al. 2003, 2004, 2005). For family 92, only *GIF* exon 7 was amplified from genomic DNA by PCR. For haplotype characterization, genotyping was performed with the six microsatellite markers, *GIF* M1 (28 kb proximal of *GIF*), *GIF* M3b (intron 7), *GIF* M4 (2 kb distal), *GIF* M5 (12 kb distal), *GIF* M6 (56 kb distal),

and *GIF* M7 (83 kb distal), and analyzed using fluorescent-labeled primers as described (Ament et al. 2009).

Results and Discussion

The two patients in family 24 were initially thought to suffer from IGS. A two-stage Schilling test was performed, where in stage 1 the patient was given radiolabeled Cbl orally, followed by parenteral Cbl to saturate tissue absorption so that any absorbed radiolabeled Cbl was released into the urine for measurement (Schilling 1953). The patient's urine over the next 24 h was negative for radiolabeled Cbl, indicative of intestinal malabsorption. For stage 2, the test was repeated by adding oral IF but this did not correct the malabsorption as would be expected for IFD. Consequently, the defect was thought to be in the intestinal cubam receptor, thereby causing IGS (Fyfe et al. 2004). As a result, the molecular genetics analysis initially focused on screening *CUBN* and *AMN* in the two patients of family 24. However, we quickly realized that the genotypes of several polymorphisms in both genes differed in the two brothers, thus excluding *CUBN* and *AMN* as the cause for this recessive disease. Given our previous experience with other cases of IGS that turned out to suffer from *GIF* mutations (Tanner et al. 2005), we proceeded with sequencing of *GIF*. Unfortunately, no obviously pathogenic mutation was detected. The only noteworthy change that cosegregated with the phenotype in the family was intronic, c.1073+5G>A in intron 7. Although

```
Wildtype Sequence Amplicon GIF exon 7:
Length: 285 nucleotides.
28.8% A, 26.7% C, 18.9% G, 25.6% T, 0.0% X, 45.6% G+C

Donor splice sites, direct strand
---------------------------------
                    pos 5'->3'  phase strand  confidence  5'     exon intr on    3'
                       178        0      +        0.71      GACCATCAAT^GTTAGTGTGA
                       195        2      +        0.00      TGAAAAGTGG^GTCAGTGTTA
Native donor site      249        2      +        0.00      CTATGTTCAA^GTGAGTATTA


Mutated Sequence Amplicon GIF exon 7:
Length: 285 nucleotides.
29.1% A, 26.7% C, 18.6% G, 25.6% T, 0.0% X, 45.3% G+C

Donor splice sites, direct strand
---------------------------------
                    pos 5'->3'  phase strand  confidence  5'     exon intron    3'
                       178        0      +        0.71      GACCATCAAT^GTTAGTGTGA
                       195        2      +        0.00      TGAAAAGTGG^GTCAGTGTTA
```

Fig. 2 Splice site prediction with NetGene2 of the 195bp amplicon covering exon 7 of GIF and the flanking splice site sequences. The upper panel shows the prediction using the wild-type sequence with the native donor splice site underlined and c.1073+5G marked in bold. The lower panel shows that the native donor splice site is predicted to be lost when c.1073+5G>A is present

it was homozygous in both patients and heterozygous in the parents and one sibling (Table 1), the molecular consequence of this change was not evident and could not be studied because of lack of RNA samples. The canonical donor splice site position +5 was G in 81.4% followed by A in 7.1% of over 22,000 intron sequences analyzed (Burset et al. 2001), thus the change was not obviously deleterious. This inconclusive result lingered while we studied additional candidate genes potentially involved in the Cbl malabsorption pathway (Shah et al. 2011).

Family 63 was referred to us for molecular analysis of suspected JCD and because of lack of proteinuria we first sequenced GIF. The same change, c.1073+5G>A in intron 7, was detected which cosegregated with the phenotype in a recessive pattern. No additional changes were identified by sequencing the complete gene. This result strongly supported that this was the pathogenic change in the patients. Since both families were of Chaldean origin, we genotyped the three flanking markers GIF M1, GIF M3b, and GIF M4 in the two families. The disease haplotype was identified as 176-mutation-235-160 in both sibships but markers GIF M1 and GIF M4 were not informative in the parents. However, the tetra-nucleotide marker GIF M3b (GGAA/GAAA with 19 different alleles observed among 93 European- and 93 African-American controls with >75% heterozygosity; Ament et al. 2009) is only 1.5 kb downstream of the mutation in intron 7. The marker was informative in three of the four parents in family 24 and 63. We concluded that GIF c.1073+5G>A is the likely culprit causing IFD in these two families because of identical haplotypes. Reverse-transcriptase PCR-analysis of GIF using RNA isolated from peripheral blood cells from family 63 failed because GIF was not expressed in these cells and stomach RNA from the patient was not obtained for ethical reasons.

Further empirical proof that c.1073+5G>A is an IFD mutation was provided by analyzing family 92. The clinical history with lack of proteinuria in the female patient supported a diagnosis of IFD. Since the family was of Iraqi Chaldean origin, we sequenced GIF exon 7 and again found c.1073+5G>A homozygous in the patient and heterozygous in the parents (Table 1). Genotyping of the patients in the three families with six microsatellite markers established a common disease haplotype. This ancestral haplotype was found unaltered in all four patients (Table 1). We concluded that the c.1073+5G>A mutation represents a founder event for IFD patients of Chaldean origin and was never seen in over 150 additional cases with Cbl malabsorption or in any of the 176 normal controls sequenced (Tanner et al. 2005). In addition, a splice site sequence prediction with NetGene2 (Brunak et al. 1991) showed that the native donor splice site is lost when c.1073+5G>A is present (Fig. 2), which strengthens the evidence that c.1073+5G>A is pathogenic. This particular change is also not listed in the dbSNP variation database. Unfortunately, Chaldean controls were not available. Iraqi Chaldeans are ethnic Assyrians. They are indigenous to, and have traditionally lived all over, Iraq, North East Syria, North West Iran, and Southeastern Turkey (Parpola 2004).

Although it could be argued that the mutation illustrated is a rare polymorphism exclusive to the Chaldean people, the phenotype of Cbl malabsorption, lack of proteinuria, and response to treatment support a defect in GIF. No other causal variant was found by sequencing most of the intronic sequences. Despite the bioinformatics results that indicate a loss of the native donor splice site (Fig. 2), the splicing defect might not remove 100% of the IF and thus some residual IF might be present. However, the range of clinical and laboratory manifestations in IFD varies considerably among patients and we do not see clear phenotype-genotype correlations. The negative Schilling test result when adding IF during stage 2 in individual 24-1 was likely due to residual general nutrient malabsorption. Lack of Cbl leads to enterocyte malfunction that in turn exacerbates Cbl uptake, leading to a vicious cycle that needs to be corrected with parenteral Cbl before any malabsorption testing is informative (Gräsbeck and Tanner 2011).

In certain cases, self-identified ethnicity to group patients for targeted genetic testing and interpretation might be advantageous to accelerate and/or confirm a diagnosis. Using this approach, we identified a novel *GIF* mutation that is responsible for Chaldean cases who suffer from inherited Cbl deficiency. Ethnic stratification also revealed the significance of a genetic variant that otherwise might have remained unexplained. The focus on personalized medicine in this example benefited a rare heritable disorder.

Acknowledgments We thank the families for supporting our research with their participation. We thank Jan Lockman and Ann-Kathrin Eisfeld for assistance in the laboratory. This work was supported by grant CA16058 from the National Cancer Institute, USA.

Declaration

The authors confirm independence from the sponsor; the content of the article has not been influenced by the sponsors. The authors have no competing financial interests.

Synopsis

Ethnicity as a shortcut to diagnosis, the case of hereditary intrinsic factor deficiency in Chaldeans.

Authors' Contributions

Amy C. Sturm and Elizabeth C. Baack coordinated DNA sample collection, genetic counseling, and helped to draft the manuscript. Michael B. Armstrong, Deborah Schiff, Ayesha Zia, and Sureyya Savasan performed clinical evaluations and sample collection. Albert de la Chapelle commented on the manuscript draft. Stephan M. Tanner carried out the molecular genetic studies, coordinated the research, and wrote the final manuscript. All authors read and approved the manuscript.

References to Electronic Databases

Online Mendelian Inheritance in Man http://omim.org

References

Ament AE, Li Z, Sturm AC et al (2009) Juvenile cobalamin deficiency in individuals of African ancestry is caused by a founder mutation in the intrinsic factor gene GIF. Br J Haematol 144:622–624

Aminoff M, Carter JE, Chadwick RB et al (1999) Mutations in CUBN, encoding the intrinsic factor-vitamin B12 receptor, cubilin, cause hereditary megaloblastic anaemia 1. Nat Genet 21:309–313

Brunak S, Engelbrecht J, Knudsen S (1991) Prediction of human mRNA donor and acceptor sites from the DNA sequence. J Mol Biol 220:49–65

Burset M, Seledtsov IA, Solovyev VV (2001) SpliceDB: database of canonical and non-canonical mammalian splice sites. Nucleic Acids Res 29:255–259

Carmel R, Green R, Rosenblatt DS, Watkins D (2003) Update on cobalamin, folate, and homocysteine. Hematology (Am Soc Hematol Educ Program):62–81. see http://www.ncbi.nlm.nih.gov/pubmed/14633777

Fyfe JC, Madsen M, Hojrup P et al (2004) The functional cobalamin (vitamin B12)-intrinsic factor receptor is a novel complex of cubilin and amnionless. Blood 103:1573–1579

Gräsbeck R (2006) Imerslund-Gräsbeck syndrome (selective vitamin B12 malabsorption with proteinuria). Orphanet J Rare Dis 1:17

Gräsbeck R, Tanner SM (2011) Juvenile selective vitamin B malabsorption: 50 years after its description – 10 years of genetic testing. Pediatr Res 70:222–228

Gräsbeck R, Gordin R, Kantero I, Kuhlbäck B (1960) Selective vitamin B12 malabsorption and proteinuria in young people. Acta Med Scand 167:289–296

Imerslund O (1960) Idiopathic chronic megaloblastic anemia in children. Acta Paediatr Scand 49:1–115

Katz M, Lee SK, Cooper BA (1972) Vitamin B 12 malabsorption due to a biologically inert intrinsic factor. N Engl J Med 287:425–429

Luder AS, Tanner SM, de la Chapelle A, Walter JH (2008) Amnionless (AMN) mutations in Imerslund-Grasbeck syndrome may be associated with disturbed vitamin B(12) transport into the CNS. J Inherit Metab Dis. doi:10.1007/s10545-007-0760-2

Obeid R, Herrmann W (2007) Holotranscobalamin in laboratory diagnosis of cobalamin deficiency compared to total cobalamin and methylmalonic acid. Clin Chem Lab Med 45:1746–1750

Ovunc B, Otto EA, Vega-Warner V et al (2011) Exome sequencing reveals cubilin mutation as a single-gene cause of proteinuria. J Am Soc Nephrol 22:1815–1820

Parpola S (2004) National and ethnic identity in the Neo-Assyrian empire and Assyrian identity in post-empire times. J Assyrian Academic Studies 18:5–22

Schilling RF (1953) Intrinsic factor studies II. The effect of gastric juice on the urinary excretion of radioactivity after oral administration of radioactive vitamin B12. J Lab Clin Med 42:860–866

Shah NP, Beech CM, Sturm AC, Tanner SM (2011) Investigation of the ABC transporter MRP1 in selected patients with presumed defects in vitamin B12 absorption. Blood 117:4397–4398

Storm T, Emma F, Verroust PJ, Hertz JM, Nielsen R, Christensen EI (2011) A patient with cubilin deficiency. N Engl J Med 364:89–91

Tanner SM, Aminoff M, Wright FA et al (2003) Amnionless, essential for mouse gastrulation, is mutated in recessive hereditary megaloblastic anemia. Nat Genet 33:426–429

Tanner SM, Li Z, Bisson R et al (2004) Genetically heterogeneous selective intestinal malabsorption of vitamin B12: founder effects, consanguinity, and high clinical awareness explain aggregations in Scandinavia and the Middle East. Hum Mutat 23:327–333

Tanner SM, Li Z, Perko JD et al (2005) Hereditary juvenile cobalamin deficiency caused by mutations in the intrinsic factor gene. Proc Natl Acad Sci U S A 102:4130–4133

Watkins D, Rosenblatt DS (2011) Inborn errors of cobalamin absorption and metabolism. Am J Med Genet C Semin Med Genet 157:33–44

Yassin F, Rothenberg SP, Rao S, Gordon MM, Alpers DH, Quadros EV (2004) Identification of a 4-base deletion in the gene in inherited intrinsic factor deficiency. Blood 103:1515–1517

JIMD Reports
DOI 10.1007/8904_2012_134

RESEARCH REPORT

Mutation Analysis in Glycogen Storage Disease Type III Patients in the Netherlands: Novel Genotype-Phenotype Relationships and Five Novel Mutations in the *AGL* Gene

Christiaan P Sentner · Yvonne J Vos ·
Klary N Niezen-Koning · Bart Mol · G Peter A. Smit

Received: 27 December 2011 / Revised: 07 February 2012 / Accepted: 13 February 2012 / Published online: 16 March 2012
© SSIEM and Springer-Verlag Berlin Heidelberg 2012

Abstract Glycogen Storage Disease type III (GSD III) is an autosomal recessive disorder in which a mutation in the *AGL* gene causes deficiency of the glycogen debranching enzyme. In childhood, it is characterized by hepatomegaly, keto-hypoglycemic episodes after short periods of fasting, and hyperlipidemia. In adulthood, myopathy, cardiomyopathy, and liver cirrhosis are the main complications. To determine the genotype of the GSD III patients ($n = 14$) diagnosed and treated in our center, mutation analysis was performed by either denaturing gradient gel electrophoresis or full gene sequencing. We developed, validated and applied both methods, and in all patients a mutation was identified on both alleles. Five novel pathogenic mutations were identified in seven patients, including four missense mutations (*c.643G>A, p.Asp215Asn; c.655A>G, p. Asn219Asp; c.1027C>T, p.Arg343Trp; c.1877A>G, p. His626Arg*) and one frameshift mutation (*c.3911delA, p.Asn1304fs*). The *c.643G>A, p.Asp215Asn* mutation is related with type IIIa, as this mutation was found homozygously in two type IIIa patients. In addition to five novel mutations, we present new genotype–phenotype relationships for *c.2039G>A, p.Trp680X; c.753_756del-

CAGA, p.Asp251fs; and the intron 32 *c.4260-12A>G* splice site mutation. The *p.Trp680X* mutation was found homozygously in four patients, presenting a mild IIIa phenotype with mild skeletal myopathy, elevated CK values, and no cardiomyopathy. The *p.Asp251fs* mutation was found homozygously in one patient presenting with a severe IIIa phenotype, with skeletal myopathy, and severe symptomatic cardiomyopathy. The *c.4260-12A>G* mutation was found heterozygously, together with the p.Arg343Trp mutation in a severe IIIb patient who developed liver cirrhosis and hepatocellular carcinoma, necessitating an orthotopic liver transplantation.

Introduction

Glycogen storage disease type III (GSD III; OMIM no. 233400) is an autosomal recessive disorder in which mutations in the *AGL* gene cause deficiency of amylo-1, 6-glucosidase and 1,4α-D-glucan 4-α-glycosyltransferase, also known as the glycogen debranching enzyme (GDE; EC no. 3.2.1.33 and 2.4.1.25). GDE catalyzes the last step in the conversion of glycogen to glucose, and GDE deficiency thus causes storage of an intermediate form of glycogen called limit dextrin (LD) (Smit et al. 2006). In the IIIa subtype, muscle and liver tissue are deficient in GDE, and this affects 85% of GSD III patients. Approximately 15% of the patients have type IIIb, in which only the liver is deficient in GDE (Shen et al. 1996). In the neonatal period and in infancy, the main features are hepatomegaly with elevated aspartate transaminase (ASAT) and alanine transferase (ALAT) values, keto-hypoglycemic episodes after relatively short periods of fasting, and hyperlipidemia.

Communicated by: Francois Feillet

Competing interests: None declared

C.P. Sentner (✉) · K.N. Niezen-Koning · G.P.A. Smit
Department of Metabolic Diseases, Beatrix Children's Hospital, University Medical Centre Groningen, University of Groningen, P.O. Box 30.001, 9700 RB Groningen, The Netherlands
e-mail: c.p.sentner@umcg.nl

Y.J. Vos · B. Mol
Department of Genetics, University Medical Centre Groningen, University of Groningen, Groningen, The Netherlands

Untreated neonates and children have developmental delay, growth retardation, and delayed puberty. In puberty and early adulthood, myopathy becomes the predominant feature of GSD III; the disease presents as a slowly progressive muscle weakness in which the proximal muscles of the shoulder and hip joints are affected. Clinical muscle weakness in the upper and the lower limb muscles can develop in later adulthood, which may be worsened by the development of peripheral neuropathy (Hobson-Webb et al. 2010; Wolfsdorf and Weinstein 2003). LD can also be stored in heart muscle, which causes a form of cardiomyopathy that resembles idiopathic hypertrophic cardiomyopathy on an echocardiogram (Lee et al. 1997; Akazawa et al. 1997).

GDE is composed of 1,532 amino acid residues and has two catalytic centers (Bao et al. 1996; Liu et al. 1991). Before GDE starts to act, a phosphorylase enzyme separates four glucose molecules from the glycogen molecule to form LD (Newgard et al. 1989). Then the first catalytic center of GDE, 1,4-glucan-4-D-glucosyltransferase, transports the three outer glucose molecules of LD to another chain. The second catalytic center, amylo-1,6-glucosidase, then releases the last glucose molecule (Ding et al. 1990). GSD III can be diagnosed biochemically by measuring GDE activity in skin fibroblasts and/or leucocytes. GDE activity and LD content can also be measured in liver and/or muscle biopsies (Wolfsdorf and Weinstein 2003). LD content can be measured in erythrocytes.

The *AGL* gene is located on chromosome 1p22 and contains 35 exons covering 85 kb of DNA (Yang-Feng et al. 1992). The full cDNA is 7 kb with a 4,596 bp coding region. A total of six mRNA isoforms are created by alternative splicing. Isoform 1 is the major isoform and widely expressed, including in liver and muscle tissue (Bao et al. 1996). Isoforms 2, 3, and 4 are present in muscle and cardiac muscle and are formed by alternative splicing or because of the difference in transcription start points. Isoform 1 contains exons 1 and 3. Isoforms 2, 3, and 4 start with exon 2. Isoforms 1 through 4 all contain exon 3, which includes the normal initiation codon for protein translation. Exons 4 through 35 are present in all isoforms (Bao et al. 1996, 1997). The glycogen-binding domain is encoded by exons 31–34. The 1,4α-D-glucan 4-α-glycosyltransferase catalytic site is encoded by exons 6, 13, 14, and 15. The amylo-1,6-glucosidase catalytic site is encoded by exons 26 and 27 (Shen and Chen 2002).

Molecular analyses of GSD III patients have been performed in several ethnic populations, and over 100 different *AGL* mutations have been described (Goldstein et al. 2010), but new mutations are still being reported (http://www.hgmd.org). No clear genotype-phenotype relationship has been established so far, although there is a relation between mutations in exon 3 and the IIIb subtype (Shen et al. 1996; Shen and Chen 2002). It is unclear, however, what mechanism enables patients with mutations in exon 3 to retain GDE activity in muscle tissue. A possible explanation was proposed by Goldstein et al. (2010) in which the exon 3 mutation is bypassed using a downstream start codon, thus creating an isoform without the exon 3 mutations.

To determine the genotype of the GSD III patients diagnosed and treated in our center (*n* = 14), the University Medical Centre Groningen (UMCG), the Netherlands, we performed mutation analysis by one of two methods. Denaturing gradient gel electrophoresis (DGGE) was applied on eight patients, and full gene sequencing was applied on six patients. We developed, validated, and applied both methods. Here we describe five novel mutations found in seven patients and their phenotypes, and evaluate the phenotype-genotype relationships in patients with previously described mutations.

Materials and Methods

Patients: For this study we analyzed 14 patients diagnosed with GSD III (seven females and seven males). They were diagnosed enzymatically by measuring GDE activity in leukocytes, fibroblasts, and/or liver tissue, and/or muscle tissue. All the patients were diagnosed and are being treated in the UMCG. Their clinical data was collected.

Mutation analysis by DGGE: Genomic DNA was isolated from EDTA blood (Miller et al. 1988). Primers were designed on the GenBank genomic reference sequence (NW_012865) to include at least 40 bp of intronic sequence on the front end of every exon and at least 20 bp at the back end. Analysis with NGRL Manchester's SNP database (http://ngrl.manchester.ac.uk/SNPCheckV2/snpcheck.htm) confirmed that no known single nucleotide polymorphisms were situated under the primers. The primer specificity was checked and verified in complete genomic sequences with NCBI's ePCR (http://www.ncbi.nlm.nih.gov/projects/e-pcr). We developed 50 primer sets to analyze 30 of the 33 coding exons by DGGE (primer sequences available upon request). The three remaining exons were sequenced directly, as designing DGGE primers with an appropriate melting curve was not possible. Per amplicon, the amplification mix consisted of 1.0 μl genomic DNA (40 ng/μl), 10 μl Amplitaq Gold ® Fast PCR Master Mix (Applied Biosystems, California, USA), 3.0 μl primer (3 pmol/μl), and 6.0 μl milliQ water; the reaction volume per sample was 20.0 μl. The samples were amplified by PCR in 96-well plates on a 96-well Gene Amp® 9700 PCR system (Applied Biosystems, California, USA).The PCR program used was as

Table 1 Sequences of the primer set designed according to the criteria of primer design used in conformation sensitive capillary electrophoresis (CSCE)[a]

Amplicon	Forward primer sequence 5′-3′	Reverse primer sequence 5′-3′	Size bp
3	CGAACATGTAAGTGCCGCTGTCA	AGAACACAGCACCATCTTTGCACAA	380
4	GTAGTGCCAAAACAGCATTAGGTTTGC	GCACTGCCATGGTTCATACAGTAACAT	457
5	TTCCATTAAGTTTTGTTGCAAC	CTGCAATGAGAGAATGGACTAATACAC	435
6	TGAACCCAAGTGTTTGACCTCTTTTCC	CCTTTCTCTTATTTGTGTGTATATGTG	432
7	AACTTTTCCTGTAACAGTATCATCG	AATACAGGTTCTAAGTAATTTTCAACC	429
8	GCACTTTGGCGTTTCTCCTGTGA	GACGTTACCCAAAAGAGAGTTTTCCCT	460
9	GGGAGGAGGTAGGAGGATAC	CACATATAGAAACATGGCCCACACACA	456
10	CTGTGTGTGGGCCATGTTTCTATATGT	TTCCCAAAAGGCAATTAACTGCCTGAA	409
11	CTGCATTTCTCCATCTGCTCTAGCAA	ATTTAAGAAATGTACTGAACTCACATG	440
12	CATCCTGCTAGATTTACTCAAAAAGCC	ACCAATAGACTAATGGGGAAGAAAATC	432
13	TTAAAAACCAGTGTTTCCTTGAAG	AATGCTTGTGTCCAACTAGC	381
14	TATGTCAAATCATGCCTCCTTTTGTC	GAAATGAGGTATCTTACCCCAAAGTAG	428
15	CCATTTCTCCAGTTAAGTTATGGG	TGGGTATGATTGTGACCAAGTGTCAGA	445
16	GGTCACAATCATACCCATATACTTC	AAACCACTGAAATCTGGACAAAGG	442
17	CTATGGCATGTTGTGCTAGTGGAAGT	TCCACATACACCTGAGAAGCAGAAAGA	433
18	AGGAGCTTGGAGCCAAGGGTTT	CCATCATACCTGGCCAAGTTACCAAA	447
19	GATTTGAAACCACTTTAGCCTTCC	TGTGGCAACTCCAGCTTGTTTAAC	340
20	TGGGACTCTCATCTTACTACTGTG	GCATGTGGATCAAGACTAACTCTG	340
22	TTGAAAACTTGTCTCCAGGAAGTG	TGGACCGTACTTTGAGTAGCAAGGAT	402
22	GAATGCTGAGTTCCTAAAACATACAC	TGCAACCCAAGTAGGCATACTCTGA	366
23	TTGTGGACTGGGTAGCCCTTGT	GAAGGAAGGAGGAAAATGGTTCAGGTT	397
24	CCTCCTTCCTTCATCATCTTTCAG	CTATCCACCTACAAGCCTTTTCAG	413
25	TGGGTGAAATGAAAGCAGTTTTG	AAAATCTTGAGTAGCATTACAAGC	458
26	ACCCCAGGTTTAGAGTAACTGTTC	CTACCTAAAGAAAATACAGCTCCC	323
27	CAAAAGTGACTGGTTTTTGTCTTC	GGTGCCAAATCAATACTGACATTTG	440
28	CTGGCCTCACCCCAATTCCTATTTC	ATTATATCGTGAGGTTTGGCACAC	352
29	CAAACTGAGCTTTAGAGTGGTTGTCCT	AGATGAAGGGAAGAAGGCAGGGAAAT	398
30	TTCATTACAATTGTTTACCGAATGCCC	GGGTTTTCCGATATTAGCTGATAG	301
31	CACATCTCAATTCAGACTGGCCACAT	AACAAATGGGAATAAGGAACTAAGC	441
32	GGCTTTCCTAACTTCTACGGCCAAAA	AGATGGCATCTCCTTTTGTTGCCC	407
33	TGCCGAGCTTATTCTGTAGAAGAC	AGGCCACAGCCACTCCTAAAAAAG	333
34	TCACCAAGGACCTGTAAGAATTTC	CCTAGGGCATACAGAAATCAATTC	350
35	CACTAGAAGGCAAAAATCACCAGGTCT	AACTTGAGCCTGTGCATATAAGGCATT	294

[a] Primer and amplicon criteria: Optimal primer length 20bp (minimum 18bp; maximum 27bp), optimal annealing temperature 58°C (minimum 52°C; maximum 64°C), optimal GC% 55 (maximum 70), optimal amplicon size 400bp (minimum 200bp; maximum 464bp), maximal amplicon GC% 73. A PT1 tail was added to every forward primer sequence (5′-3′; TGTAAAACGACGGCCAGT), and a PT2 tail was added to every reverse primer sequence (5′-3′; CAGGAAACAGCTATGACC)

follows: 3 min at 96°C, 45 cycles of 1 min at 96°C, 1 min annealing at multiple temperatures, 1 min elongation at 72°C, and a final extension step at 72°C lasting 5 min. The annealing temperatures were 5 cycles at 60°C, 5 cycles at 56°C, 5 cycles at 52°C, and 30 cycles at 50°C. The PCR products were analyzed by DGGE (Hayes et al. 1999). Amplicons showing an aberrant banding pattern were sequenced on an ABI 3730 automated sequencer (Applied Biosystems, California, USA) using specific primers. DNA samples from 38 GSD III patients with 34 known

mutations were used to validate the DGGE system – and all mutations were detected in our system.

Sequence analysis: We designed 33 primer sets according to the criteria used in conformation-sensitive capillary electrophoresis (CSCE, Table 1) and applied them for sequencing all coding exons of the gene including flanking intronic sequences. Per sample, the amplification mix consisted of 2 μl genomic DNA, 5 μl Amplitaq Gold® Fast PCR Master Mix (Applied Biosystems, California, USA), and 3.0 μl primer (150 fmol/μl). The reaction volume per sample was 10 μl; the samples were amplified by PCR in 384-well plates on a Veriti 384-well thermal cycler (Applied Biosystems, California, USA). The PCR program was as follows: 3 min at 94°C, 35 cycles of 1 min at 94°C, 1 min annealing at 60°C, 1 min at 72°C, and a final extension step at 72°C lasting 7 min, after which the samples were cooled down to 20°C. Five microliters of the PCR products were loaded with 5 μl loading buffer and run on a 2% agarose gel with a FastRuler Low Range DNA ladder (Fermentas, Vilnius, Lithuania) for comparison. The remaining PCR products were purified with ExoSAP-IT (Amersham Pharmacia Biotech, Piscataway, NY, USA) and subjected to direct sequencing on an ABI 3730 automated sequencer (Applied Biosystems, California, USA) using specific primers.

Analysis of the pathogenicity of novel mutations: The pathogenicity of novel mutations was assessed using six separate methods. One hundred control chromosomes from mixed ethnicity were checked for novel mutations. Conservation of the mutated nucleotide and amino acid was graded with Alamut Version 1.4 (©Interactive Biosoftware). The University of Harvard's PolyPhen program (http://genetics.bwh.harvard.edu/pph/) predicted the impact of an amino acid substitution on the structure and function of a human protein. The SIFT program (http://blocks.fhcrc.org/sift/SIFT.html) predicted whether an amino acid substitution affects protein function based on sequence homology and the physical properties of amino acids. Finally, we assessed whether the mutation was located in an exon encoding the glycogen binding domain (exon 31-34), or encoding a catalytic site (exon 6, 14 16, 26-27), and measured the GDE activity.

Results

Patients: Of our 14 GSD III patients, 4 were related to one other patient (two sisters, and two brothers), all other patients were unrelated. Patient 7 was born to consanguineous parents. Nine patients had type IIIa, two patients had type IIIb, and three pediatric patients were too young to be subtyped based on their clinical presentation. The clinical and biochemical characteristics of the patients are presented in Table 2.

Results of mutation analysis: The patients were fully analyzed and we detected two mutations in each patient (Table 3). Five novel mutations were identified in seven patients, including two sisters who were homozygous for *c.643G>A, p.Asp215Asn* and two brothers who were homozygous for *c.3911delA, p.Asn1304fs*. The other three patients were compound heterozygous with a novel missense mutation and a previously reported pathogenic mutation: *c.655A>G, p.Asn219Asp* in combination with

c.4529dupA, p.Tyr1510X; c.1027C>T, p.Arg343Trp in combination with the splice mutation *c.4260-12A>G*; and *c.1877A>G, p.His626Arg* in combination with *c.1222C>T, p.Arg408X.*

Results of pathogenicity analysis: We consider the four new missense mutations to be pathogenic for several reasons. We did not detect the new mutations in 100 control chromosomes, and these mutations were not found in the NCBI SNP database. Highly conserved amino acids were mutated, and all mutations were predicted to affect protein function by the PolyPhen and SIFT programs. Additionally, three of the four new mutations (*Asp215Asn, p.Asn219Asp*, and *p.His626Arg*) were located in an exon encoding the 1,4α-D-glucan 4-α-glycosyltransferase catalytic site of GDE (Fig. 1). This makes a pathogenic effect of these mutations on GDE function very probable (Cheng et al. 2009). As the *c.3911delA, p.Asn1304fs* causes a frameshift with the new reading frame ending in a stop codon at position 10, this is considered to be a pathogenic mutation as well. Clinically, all the patients presented with a GSD III phenotype, and this was biochemically confirmed by enzymatic analysis that no GDE activity was measured to be present.

Discussion

Here we describe five novel pathogenic mutations in the *AGL* gene, four of which are missense mutations, very likely causing GSD III in seven patients. Missense mutations causing GSD III are scarce, as truncating mutations compose the majority of the pathogenic mutations. However, pathogenic missense mutations have been described. Cheng et al. showed that missense mutations located in the active sites produce a GSD III phenotype in which there is total or partial abolishment of GDE activity depending on the location of the mutation (Cheng et al. 2009).

GSD III is characterized by clinical and genotypical heterogeneity, and clear genotype-phenotype correlations are rare. We therefore present the clinical characteristics of the patients with novel mutations, and discuss the correspondence between their clinical presentation and that described in the literature for the mutations that have been previously described.

Genotype-phenotype analysis: Patients 1 to 4 all had type IIIa and were homozygous for *p.Trp680X*. The phenotype was similar in all four patients, with hepatomegaly, no or mild cardiac involvement, and mild skeletal myopathy with elevated CK values. This suggests a link between the *p.Trp680X* mutation and the type IIIa phenotype. The *p.Trp680X* mutation was described previously in a compound heterozygote patient, in whom it was linked to

Table 2 Demographic, clinical, and biochemical characteristics of the analyzed GSD III patients

Patient no.	Age (years)	Sex	Subtype	Ethnic origin	GDE residual activity (%)	Liver complications	Cardiologic complications	Skeletal muscle complications	Most recent CK value (U/L)
1	8	M	IIIa	Caribbean	0	Hepatomegaly	None	Proximal myopathy	1083
2	25	F	IIIa	Caribbean	0	Hepatomegaly	None	None	2232
3	26	M	IIIa	Caribbean	0	Hepatomegaly	Septal hypertrophy	None	893
4	40	F	IIIa	Caribbean	0	Hepatomegaly	None	None	3729
5	15	F	IIIa	Mediterranean	0	Hepatomegaly	None	None	662
6	20	F	IIIa	Mediterranean	0	Hepatomegaly	None	None	1342
7	32	F	IIIa	Mediterranean	0	Hepatomegaly	Severe symptomatic left ventricular and septal hypertrophy	Exercise intolerance, distal myopathy	1898
8	3	M	Too young	Caucasian	1	Hepatomegaly	None	None	133
9	30	F	IIIa	Caucasian	0	None	None	Exercise intolerance, distal myopathy	2257
10	30	M	IIIb	Caucasian	0	Hepatomegaly	None	None	68
11	41	F	IIIa	Caucasian	0	Hepatomegaly	None	Exercise intolerance	392
12	41	F	IIIb	Caucasian	0	None, the patient is post-OLT[c],pre-OLT[c] liver cirrhosis and hepatocellular carcinoma was present	None	None	70
13	3	M	Too young	Caucasian	0	Hepatomegaly	None	None	128
14	1	M	Too young	Caucasian	0	Hepatomegaly	None	None	466

M Male, *F* Female; *OLT* Orthotopic Liver Transplantation

the IIIb phenotype (Shen et al. 1996), which is in contrast to what we found in our patients. Interestingly, all our patients with the *c.2039G>A, p.Trp680X* mutation were from the same topographic region and ethnic origin, making a founder effect feasible, as seen for the *c.1222C>T, p. Arg408X* mutation on the Faroe Islands (Santer et al. 2001). However, in order to prove this, haplotyping for these patients would be required, which we have not done in this study.

Patient 5 (female, age 15 years) and patient 6 (female, age 20 years) were sisters (whose parents were from the Mediterranean) who had type IIIa, and were homozygous for the novel *p.Asp215Asn* mutation. Their phenotype was mild with hepatomegaly as the main finding, there was no myopathy or cardiomyopathy. In laboratory investigations ASAT, ALAT, and CK values were elevated, but there was no hypoglycemia or hyperlipidemia. The mild phenotype

was probably due to their young age and the good metabolic control of these patients.

Patient 7 (female, age 32 years) had type IIIa and was found to be homozygous for *p.Asp251fs*. She had a severe IIIa phenotype, with hepatomegaly, and skeletal muscle involvement with severe exercise intolerance. There was also severe cardiac involvement with hypertrophy of the interventricular septum and left ventricle, necessitating pharmacological treatment and ICD placement. This mutation was previously described in a 3-year-old female, who was found to have hepatosplenomegaly and hypoglycemia but no myopathy or cardiomyopathy (Lucchiari et al. 2006).

In patient 8 (male, age 3 years), the novel mutation *p.Asn219Asp* was found to be heterozygous. The other mutation *p.Tyr1510X* was previously reported and associated with a severe IIIa phenotype (Shen et al. 1997). At first

Table 3 Mutation analysis results in 14 GSD III patients

Patient no.	Exon no.	Nucleotide change allele 1	Amino acid change allele 1	Mutation type	Exon no.	Nucleotide change allele 2	Amino acid change allele 2	Mutation type	Mutation analysis method
1	17	c.2039G>A	p.Trp680X	Nonsense	17	c.2039G>A	p.Trp680X	Nonsense	DGGE
2	17	c.2039G>A	p.Trp680X	Nonsense	17	c.2039G>A	p.Trp680X	Nonsense	DGGE
3	17	c.2039G>A	p.Trp680X	Nonsense	17	c.2039G>A	p.Trp680X	Nonsense	DGGE
4	17	c.2039G>A	p.Trp680X	Nonsense	17	c.2039G>A	p.Trp680X	Nonsense	Sequencing
5	6	c.643G>A	p.Asp215Asn	Missense	6	c.643G>A	p.Asp215Asn	Missense	DGGE
6	6	c.643G>A	p.Asp215Asn	Missense	6	c.643G>A	p.Asp215Asn	Missense	DGGE
7	7	c.753_756delCAGA	p.Asp251fs	Frameshift	7	c.753_756delCAGA	p.Asp251fs	Frameshift	DGGE
8	6	c.655A>G	p.Asn219Asp	Missense	35	c.4529dupA	p.Tyr1510X	Nonsense	Sequencing
9	35	c.4529dupA	p.Tyr1510X	Nonsense	35	c.4529dupA	p.Tyr1510X	Nonsense	DGGE
10	3	c.16C>T	p.Gln6X	Nonsense	3	c.16C>T	p.Gln6X	Nonsense	DGGE
11	11	c.1222C>T	p.Arg408X	Nonsense	15	c.1877A>G	p.His626Arg	Missense	Sequencing
12	9	c.1027C>T	p.Arg343Trp	Missense	Intron 32	c.4260-12A>G		Splice	Sequencing
13	30	c.3911delA	p.Asn1304fs	Frameshift	30	c.3911delA	p.Asn1304fs	Frameshift	Sequencing
14	30	c.3911delA	p.Asn1304fs	Frameshift	30	c.3911delA	p.Asn1304fs	Frameshift	Sequencing

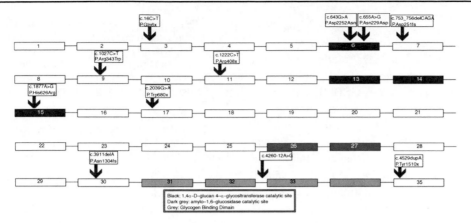

Fig. 1 The location of all found mutations in the gene, the catalytic sites of GDE are depicted as well; three of the four novel mutations (*Asp215Asn*; *p.Asn219Asp*; and *p.His626Arg*) were located in exons encoding the 1,4α-D-glucan 4-α-glycosyltransferase catalytic site

presentation (age 1.5 years), he had severe hepatomegaly extending 15 cm below the costal margin in the medioclavicular line, regular keto-hypoglycemic episodes and hyperlipidemia. These symptoms improved dramatically upon starting dietary treatment after diagnosis at the age of 1.5 years with frequent meals during the day and overnight gastric drip feeding. These findings are typical for GSD III patients in this age group, and the normal CK value does not exclude future muscle involvement, making it hard to assess the subtype or to determine the phenotype that goes with this novel mutation. Furthermore, the latter proves difficult because the patient was found to have compound heterozygous mutations.

Patient 9 (female, 30 years) had type IIIa and was found to be homozygous for *p.Tyr1510X*. She had a severe IIIa phenotype including distal myopathy and severe exercise intolerance with elevated CK (2257 U/L). Despite the absence of hepatomegaly, transaminases remained elevated (ASAT 237 U/L, ALAT 175 U/L). This is in concordance with a suggestion made by Shen et al. that this mutation is associated with a severe GSD IIIa phenotype (Shen et al. 1997).

Patient 10 (male, 30 years) had type IIIb and was homozygous for *p.Gln6X*. His phenotype was mild, with hepatomegaly, normal CK values, and no cardiac or skeletal muscle involvement. This mutation was previously described and is strongly linked with the GSD IIIb phenotype, as are other mutations in exon 3 (Shen et al. 1996; Shen and Chen 2002). Paradoxically, this patient first presented at the age of 2 years with proximal myopathy and severe hypotonia, but without cardiomyopathy or elevated CK values. His myopathy and hypotonia improved dramatically upon starting dietary treatment, and he has had no muscular symptoms since.

Patient 11 (female, age 41 years) had type IIIa and was compound heterozygous for the novel *p.His626Arg* mutation. Her other mutation, *p.Arg408X*, was previously described

(Santer et al. 2001; Lam et al. 2004) and associated with the GSD IIIa phenotype. This corresponds with her phenotype: hepatomegaly and elevated transaminases. She has no proximal or distal myopathy but she does suffer from exercise intolerance and has elevated CK values.

Patient 12 (female, age 41 years) had type IIIb and was compound heterozygous for the novel *p.Arg343Trp* mutation and *c.4260-12A>G* in intron 32. *c.4260-12A>G* was previously described and associated with IIIb as well as IIIa (Okubo et al. 1998; Shaiu et al. 2000). This patients' phenotype was severe IIIb and a case report on her was published after she received an orthotopic liver transplantation after being diagnosed with liver cirrhosis (Haagsma et al. 1997). Hepatocellular carcinoma was found upon pathological examination of the excised liver. She has never had any muscle involvement, and CK values have always been normal.

Patients 13 (male, age 3 years) and 14 (male, age 1 years) were brothers and homozygous for the novel mutation *c.3911delA*, *p.Asn1304fs* in exon 30. Except for prominent hepatomegaly and the need for appropriate dietary requirements, both patients were generally well. There was no clinical muscle involvement, even though patient 14 had elevated CK values. As these are very young pediatric patients, it is not yet possible to assess the subtype based on the clinical findings.

Conclusions: We identified two separate mutations in each of our 14 GSD III patients. Five were novel pathogenic mutations considered to be causal. As we also analyzed parts of the intron sequences (40 bp on the front end, and 20 bp at the back end of every exon), we assume that the chance that we missed other disease-causing mutations is slim. The novel *c.643G>A*, *p.Asp215Asn* mutation is related with type IIIa, as this mutation was found homozygously in two patients both clinically presenting as type IIIa. We also established new genotype-phenotype relationships between *c.2039G>A*,

p.Trp680X and type IIIa; *c.753_756delCAGA, p.Asp251fs*
and type IIIa; and the intron 32 *c.4260-12A>G* splice site
mutation and type IIIb. The association between
c.4529dupA, p.Tyr1510X and type IIIa complies with
previous literature. However, as the GSD III subtype is
not yet clear for every patient, it was not possible to
establish a genotype-phenotype relationship for every novel
mutation. There is still a large clinical and genotypical
heterogeneity in GSD III, which makes establishing
genotype-phenotype relationships for GSD III difficult.
The fact that we found five new mutations in a relatively
small number of GSD III patients further accentuates the
need for more genotyping, and indicates that there are
probably numerous unidentified mutations.

Acknowledgments   We thank Prof. René Santer of the University
Clinic Hamburg, Germany, for providing the DNA samples for the
validation of our methods.

Synopsis

GSD III is characterized by a large clinical and genotypical
heterogeneity, which makes the establishment of clear
genotype-phenotype relationships difficult.

Conflicts of Interest

None declared.

References

Akazawa H, Kuroda T, Kim S, Mito H, Kojo T, Shimada K (1997)
    Specific heart muscle disease associated with glycogen storage
    disease type III: clinical similarity to the dilated phase of
    hypertrophic cardiomyopathy. Eur Heart J 18:532–533
Bao Y, Dawson TL Jr, Chen YT (1996) Human glycogen debranching
    enzyme gene (*AGL*): complete structural organization and
    characterization of the 5' flanking region. Genomics 38:155–165
Bao Y, Yang BZ, Dawson TL Jr, Chen YT (1997) Isolation and
    nucleotide sequence of human liver glycogen debranching
    enzyme mRNA: identification of multiple tissue-specific iso-
    forms. Gene 15:389–398
Cheng A, Zhang M, Okubo M, Omichi K, Saltiel AR (2009) Distinct
    mutations in the glycogen debranching enzyme found in
    glycogen storage disease type III lead to impairment in diverse
    cellular functions. Hum Mol Genet 18:2045–2052
Ding JH, De Barsy T, Brown BI, Coleman RA, Chen YT (1990)
    Immunoblot analyses of glycogen debranching enzyme in
    different subtypes of glycogen storage disease type III. J Pediatr
    116:95–100
Goldstein JL, Austin SL, Boyette K, Kanaly A, Veerapandiyan A,
    Rehder C, Kishnani PS, Bali DS (2010) Molecular analysis of the
    AGL gene: identification of 25 novel mutations and evidence of
    genetic heterogeneity in patients with glycogen storage disease
    type III. Genet Med 12:424–430
Haagsma EB, Smit GPA, Niezen-Koning KE, Gouw AS, Meerman L,
    Slooff MJ (1997) Type IIIb glycogen storage disease associated
    with end-stage cirrhosis and hepatocellular carcinoma. The Liver
    Transplant Group. Hepatology 25:537–540
Hayes VM, Wu Y, Osinga J, Mulder IM, van der Vlies P, Elfferich P,
    Buys CH, Hofstra RM (1999) Improvements in gel composition
    and electrophoretic conditions for broad-range mutation analysis
    by denaturing gradient gel electrophoresis. Nucleic Acids Res 27:
    e29
Hobson-Webb LD, Austin SL, Bali DS, Kishnani PS (2010) The
    electrodiagnostic characteristics of glycogen storage disease type
    III. Genet Med 12:440–445
Lam CW, Lee AT, Lam YY, Wong TW, Mak TW, Fung WC, Chan
    KC, Ho CS, Tong SF (2004) DNA-based subtyping of glycogen
    storage disease type III: mutation and haplotype analysis of the
    AGL gene in Chinese. Mol Genet Metab 83:271–275
Lee PJ, Deanfield JE, Burch M, Baig K, McKenna WJ, Leonard JV
    (1997) Comparison of the functional significance of left
    ventricular hypertrophy in hypertrophic cardiomyopathy and
    glycogenosis type III. Am J Cardiol 79:834–838
Liu W, Madsen NB, Braun C, Withers SG (1991) Reassessment of the
    catalytic mechanism of glycogen debranching enzyme. Biochem-
    istry 30:1419–1424
Lucchiari S, Pagliarani S, Salani S, Filocamo M, Di Rocco M, Melis
    D, Rodolico C, Musumeci O, Toscano A, Bresolin N, Comi GP
    (2006) Hepatic and neuromuscular forms of glycogenosis type
    III: nine mutations in *AGL*. Hum Mutat 27:600–601
Miller SA, Dykes DD, Polesky HF (1988) A simple salting out
    procedure for extracting DNA from human nucleated cells.
    Nucleic Acids Res 16:1225
Newgard CB, Hwang PK, Fletterick RJ (1989) The family of
    glycogen phosphorylases: structure and function. Crit Rev
    Biochem Mol Biol 24:69–99
Okubo M, Horinishi A, Nakamura N, Aoyama Y, Hashimoto M, Endo
    Y, Murase T (1998) A novel point mutation in an acceptor splice
    site of intron 32 (IVS32 A-12→G) but no exon 3 mutations in
    the glycogen debranching enzyme gene in a homozygous patient
    with glycogen storage disease type IIIb. Hum Genet 102:1–5
Santer R, Kinner M, Steuerwald U, Kjaergaard S, Skovby F,
    Simonsen H, Shaiu WL, Chen YT, Schneppenheim R, Schaub J
    (2001) Molecular genetic basis and prevalence of glycogen
    storage disease type IIIa in the Faroe Islands. Eur J Hum Genet
    9:388–391
Shaiu WL, Kishnani PS, Shen J, Liu HM, Chen YT (2000) Genotype-
    phenotype correlation in two frequent mutations and mutation
    update in type III glycogen storage disease. Mol Genet Metab
    69:16–23
Shen J, Bao Y, Liu HM, Lee PJ, Leonard JV, Chen YT (1996)
    Mutations in Exon 3 of the glycogen debranching enzyme gene
    are associated with glycogen storage disease type III that is
    differentially expressed in liver and muscle. J Clin Invest
    98:352–357
Shen J, Bao Y, Chen YT (1997) A nonsense mutation due to a single
    base insertion in the 3'-coding region of glycogen debranching
    enzyme gene associated with a severe phenotype in a patient with
    glycogen storage disease type IIIa. Hum Mutat 9:37–40
Shen JJ, Chen YT (2002) Molecular characterization of glycogen
    storage disease type III. Curr Mol Med 2:167–175
Smit GPA, Rake JP, Akman HO, DiMauro S (2006) The glycogen
    storage diseases and related disorders. In: Fernandes J, Saudu-
    bray J-M, van den Berghe G, Walter JH (eds) Inborn metabolic
    diseases: diagnosis and treatment. Springer Medizin Verlag,
    Heidelberg, pp 103–116
Wolfsdorf JI, Weinstein DA (2003) Glycogen storage diseases. Rev
    Endocr Metab Disord 4:95–102
Yang-Feng TL, Zheng K, Yu J, Yang BZ, Chen YT, Kao FT (1992)
    Assignment of the human glycogen debrancher gene to chromo-
    some 1p22. Genomics 13:931–934

JIMD Reports
DOI 10.1007/8904_2012_135

CASE REPORT

Cholestatic Jaundice Associated with Carnitine Palmitoyltransferase IA Deficiency

AAM Morris · SE Olpin · MJ Bennett · A Santani ·
J Stahlschmidt · P McClean

Received: 09 February 2012 / Revised: 09 February 2012 / Accepted: 15 February 2012 / Published online: 21 March 2012
© SSIEM and Springer-Verlag Berlin Heidelberg 2012

Abstract Liver dysfunction usually accompanies metabolic decompensation in fatty acid oxidation disorders, including carnitine palmitoyltransferase (CPT) Ia deficiency. Typically, the liver is enlarged with raised plasma transaminase activities and steatosis on histological examination. In contrast, cholestatic jaundice is rare, having only been reported in long-chain 3-hydroxyacyl-CoA dehydrogenase (LCHAD) deficiency. We report a 3-year-old boy with CPT Ia deficiency who developed hepatomegaly and cholestatic jaundice following a viral illness.

Communicated by: Rodney Pollitt

Competing interests: None declared

A.A.M. Morris (✉)
Willink Unit, Genetic Medicine, Manchester Academic Health
Sciences Centre, Central Manchester University Hospitals NHS
Foundation Trust, Oxford Road,
Manchester M13 9WL, UK
e-mail: Andrew.morris@cmft.nhs.uk

S.E. Olpin
Department of Clinical Chemistry, Sheffield Children's Hospital,
Sheffield, UK

M.J. Bennett
Metabolic Disease, The Children's Hospital of Philadelphia,
Philadelphia, USA

A. Santani
Molecular Diagnostics Laboratories, The Children's Hospital
of Philadelphia, Philadelphia, USA

J. Stahlschmidt
Department of Histopathology, Leeds Teaching Hospitals
NHS Trust, Leeds, UK

P. McClean
Children's Liver Unit, Leeds Teaching Hospitals NHS Trust,
Leeds, UK

No cause for the jaundice could be found, apart from the fatty acid oxidation disorder. Liver histology showed diffuse, predominately macrovesicular steatosis, hepatocellular and canalicular cholestasis but no bile duct paucity or evidence of large duct obstruction. The liver dysfunction resolved in 4–7 weeks.

Abbreviations

CMV	Cytomegalovirus
CPT	Carnitine palmitoyltransferase
EBV	Epstein-Barr virus
LCHAD	Long-chain 3-hydroxyacyl-CoA dehydrogenase

Introduction

Long-chain fatty acids are activated to coenzyme A (CoA) esters at the outer mitochondrial membrane but they need to be converted to carnitine esters in order to cross the inner mitochondrial membrane. This reaction is catalysed by carnitine palmitoyltransferase I (CPT I, EC 2.3.1.21). Within the mitochondria, the acyl-groups are converted back to CoA esters by CPT II. Three different genetic isoforms of CPT I have been found in different tissues – CPT Ia in liver and kidney, CPT Ib in muscle and heart and CPT Ic in brain.

Of the three isoforms, only CPT Ia deficiency has been identified in humans (OMIM 255120). Most patients present by the age of 2 years with hypoketotic hypoglycaemia, induced by fasting or illness. This is usually accompanied by liver dysfunction; there may also be transient lipaemia and renal tubular acidosis (Olpin et al. 2001). Paradoxically, although CPT Ia is not highly expressed in heart or muscle, several patients have had

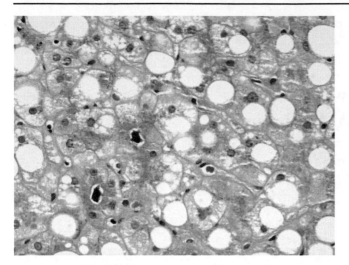

Fig. 1 Liver histology showing steatosis (predominantly macro-vesicular) and cholestasis. H&E × 400

cardiac problems or raised plasma creatine kinase values (Olpin et al. 2001).

The liver dysfunction that accompanies hypoglycaemia in CPT Ia deficiency resembles that seen in other fatty acid oxidation disorders, such as MCAD deficiency, but it may persist for longer, often for several weeks. The liver is enlarged and histology has shown steatosis (Bougneres et al. 1981). Plasma transaminase activities are increased and there may be mild hyperammonaemia and hyperuricaemia. Here we report a patient who showed a different pattern of liver dysfunction, with cholestasis.

Case Report

The patient is the only child of healthy, non-consanguineous Caucasian parents. He presented at 18 months of age with a respiratory arrest and hypoglycaemia (0.4 mmol/l), associated with gastroenteritis. He made a full recovery but, at 2 years of age, a further episode of gastroenteritis led to a 5-min seizure associated with hypoglycaemia (2.1 mmol/l). Following correction of the hypoglycaemia, the patient had some brief focal seizures and was found to have left-sided weakness; over 24 h, he developed marked hepatomegaly. There was lipaemia (triglycerides 36 mmol/l, normal <1.7) with raised plasma levels of creatine kinase (1806 IU/l, normal 24–195) and urate (1088 umol/l, normal 120–390). Echocardiography, electrocardiography and cranial MRI were all normal. Urine organic acid analysis was unremarkable, as was blood acylcarnitine analysis, except for a raised free carnitine concentration (82 umol/l, normal 14–74). Tritium release assays subsequently showed reduced fatty acid oxidation flux (17–21% simultaneous controls for myristate, 12% for palmitate and 9–13% for oleate).

Acylcarnitine measurements in cultured fibroblasts after 96 h incubation with 400 umol/l carnitine and 200 umol/l palmitate showed a low C16-acylcarnitine (0.26 umol/l, controls 1.47 ± 0.76, mean ± 2SD, $n = 65$) and a raised C5/C16 acylcarnitine ratio (7.3, controls 0.13–1.01, $n = 70$). CPT1a mutation analysis showed that the patient was a compound heterozygote for c.1766_1767insACATA (p.Tyr589X) and c.526G>T (p.Val176Phe). The patient made a full recovery, the hepatomegaly and weakness resolving over 6–8 weeks. The family was advised to give regular drinks containing glucose polymer during illnesses.

At the age of 3 years 9 months, the patient had a further admission to hospital. He had cough, diarrhoea, vomiting and fever, persisting for 6 days. There was an outbreak of influenza A H1N1 in the area and his general practitioner treated him with a course of oseltamivir, although the diagnosis was not confirmed by microbiological techniques. Glucose polymer was not given as he continued to eat. On admission, he was not hypoglycaemic but he had hepatomegaly and deranged liver function tests (bilirubin 79 μmol/l, alanine transaminase 336 IU/l, gamma glutamyltransferase 153 IU/l and normal clotting parameters). His symptoms settled after 2–3 days and he was allowed home with outpatient follow-up. After 10 days, he was readmitted with increasing jaundice, marked hepatomegaly and mild dehydration. He had no stigmata of chronic liver disease and was not encephalopathic. Initially, blood was lipaemic (triglycerides 12.2.mmol/l). Blood glucose concentrations were consistently normal. The plasma bilirubin concentration had increased to 226 μmol/l and stayed at this level for the next 7 days. Plasma activities were raised for alanine transaminase (maximum value 446 IU/l normal 9–36), alkaline phosphatase (maximum 826 IU/l normal 60–370) and gamma glutamyltransferase (maximum 368 IU/l normal <50). The albumin concentration was normal as were clotting studies apart from a marginally prolonged prothrombin time (14 s, controls 9–12 s).

No cause for the cholestatic liver disease was identified, apart from the fatty acid oxidation disorder. Serum alpha-1-antitrypsin, caeruloplasmin and immunoglobulins were normal and an autoantibody screen and serology for hepatitis A, B, C, cytomegalovirus (CMV), Epstein-Barr virus (EBV) and Varicella were negative. Polymerise chain reactions for CMV, EBV, entero- and adeno-viruses were also negative. Abdominal ultrasound examination showed an enlarged, echobright liver with no biliary dilatation; the spleen was normal, as was flow in the hepatic artery, hepatic veins and hepatic portal vein. A liver biopsy, undertaken 4 weeks after the initial admission, showed diffuse macrovesicular steatosis with minor microvesicular change in 60–80% hepatocytes. There was hepatocellular and canalicular cholestasis with focal acinar formation but no bile duct paucity, no evidence of large duct obstruction and no significant fibrosis (Fig. 1). Copper-associated protein was not identified.

The patient was given a diet high in carbohydrate from the time of admission. Supplements of medium-chain triglycerides, fat-soluble vitamins and ursodeoxycholic acid were added subsequently. The liver size and liver function tests gradually returned to normal 4–7 weeks after admission.

Discussion

It seems very likely that this boy's cholestasis was caused by the metabolic derangement. Steatosis was the main finding on liver histology, as in previous patients with CPT Ia deficiency (Bougneres et al. 1981). The only additional features were hepatocellular and canalicular cholestasis, with focal acinar formation. The patient was thought to have had a viral illness, possibly influenza A H1N1, but the liver histology was not typical of a viral-induced hepatitis. Moreover, H1N1 influenza is associated with mild to moderate transaminitis but not severe cholestasis (Yingying 2011). The Medicines and Healthcare Products Regulatory Agency has received two reports of liver dysfunction after treatment with oseltamivir but there were alternative causes of liver dysfunction in these cases and a causal association with the drug was not established (www.mhra.gov.uk/home/groups/pl-p/documents/websiteresources/con059973.pdf).

Liver dysfunction is common during acute illnesses in many fatty acid oxidation disorders. In contrast, cholestatic jaundice is rare, having only been reported previously in patients with LCHAD deficiency (Ibdah et al. 1999; Saudubray et al. 1999; Tyni et al. 1997). In these patients, the cholestasis has often accompanied an acute Reye-like illness but sometimes it has been an incidental finding (e.g. in a patient presenting with hypocalcaemia). Pregnant mothers carrying a foetus with LCHAD deficiency also have an increased risk of cholestasis (though the increase in risk is less than for HELLP syndrome and acute fatty liver of pregnancy) (Tyni et al. 1998).

CPT Ia deficiency is generally rare but a variant in the gene is extremely common in the Inuit population of Alaska, Canada and Greenland. In some regions, 70% of babies are homozygous for the c.1436C>T mutation, which reduces CPT Ia activity, estimates of residual activity ranging from 6–22% of control values. A few homozygous subjects present with hypoglycaemia as neonates or young children but most have been considered asymptomatic (Greenberg et al. 2009). Interestingly, however, a number

of them have self-limiting cholestatic jaundice during infancy (S. Mercimek-Mahmutoglu, 2010, personal communication).

Our patient's cholestasis might be due to impaired energy production preventing ATP-dependent bile acid secretion; alternatively, toxic metabolites might interfere with the canalicular transport proteins for bile acids or phospholipids. Though cholestasis is unusual in fatty acid oxidation disorders, it accompanies hepatic steatosis in a number of other genetic metabolic disorders, including mitochondrial liver disease, galactosaemia, tyrosinaemia type 1, Wilson disease and cystic fibrosis. Mild cholestasis also occurs in a rat model of non-alcoholic fatty liver disease (Pizarro et al. 2004) but the pathogenesis is unlikely to resemble that in our patient.

Our patient's liver dysfunction resolved over 4–7 weeks on treatment with a high carbohydrate diet, medium chain triglycerides and ursodeoxycholic acid. We do not know whether the treatment contributed to the recovery, but we suspect that the jaundice would have resolved without intervention.

References

Bougneres PF, Saudubray JM, Marsac C, Bernard O, Odievre M, Girard J (1981) Fasting hypoglycemia resulting from hepatic carnitine palmitoyl transferase deficiency. J Pediatr 98:742–746

Greenberg CR, Dilling LA, Thompson GR et al (2009) The paradox of the carnitine palmitoyltransferase type Ia P479L variant in Canadian aboriginal populations. Mol Genet Metab 96:201–207

Ibdah JA, Dasouki MJ, Strauss AW (1999) Long-chain 3-hydroxyacyl-CoA dehydrogenase deficiency: variable expressivity of maternal illness during pregnancy and unusual presentation with infantile cholestasis and hypocalcaemia. J Inherit Metab Dis 22:811–814

Olpin SE, Allen J, Bonham JR et al (2001) Features of carnitine palmitoyltransferase type I deficiency. J Inherit Metab Dis 24:35–42

Pizarro M, Balasubramaniyan N, Solis N et al (2004) Bile secretory function in the obese Zucker rat: evidence of cholestasis and altered canalicular transport function. Gut 53:1837–1843

Saudubray JM, Martin D, de Lonlay P et al (1999) Recognition and management of fatty acid oxidation defects: a series of 107 patients. J Inherit Metab Dis 22:488–502

Tyni T, Ekholm E, Pihko H (1998) Pregnancy complications are frequent in long-chain 3-hydroxyacyl-coenzyme A dehydrogenase deficiency. Am J Obstet Gynecol 178:603–608

Tyni T, Palotie A, Viinikka L et al (1997) Long-chain 3-hydroxyacyl-coenzyme A dehydrogenase deficiency with the G1528C mutation: clinical presentation of thirteen patients. J Pediatr 130:67–76

Yingying C (2011) Abnormal liver chemistry in patients with influenza A H1N1. Liver Int 31:902

JIMD Reports
DOI 10.1007/8904_2012_136

Quality of Life of Brazilian Patients with Gaucher Disease and Fabry Disease

Fabiane Lopes Oliveira · Taciane Alegra · Alicia Dornelles · Bárbara Corrêa Krug ·
Cristina B. O. Netto · Neusa Sica da Rocha · Paulo D. Picon ·
Ida Vanessa D. Schwartz

Received: 16 June 2011 / Revised: 16 February 2012 / Accepted: 17 February 2012 / Published online: 18 April 2012
© SSIEM and Springer-Verlag Berlin Heidelberg 2012

Summary *Objective*: To evaluate QoL in a sample of Brazilian patients with Gaucher (GD) and Fabry (FD) disease using the SF-36 survey.

Method: Observational cross-sectional study. The SF-36 survey was administered to cognitively able patients 12 years or older, who were seen in the Medical Genetics Service of Hospital de Clínicas de Porto Alegre, Brazil.

Results: Thirty-five patients were included in the study (GD = 21, FD = 14), mean age was 29.8 ± 14.2 years and 29 (82.9%) were receiving ERT. Patients with GD receiving ERT had better scores in the general health ($p = 0.046$) domain of the SF-36 than patients with FD receiving ERT. Comparison of patients with GD naive to ERT and those receiving ERT revealed differences only in the bodily pain domain ($p = 0.036$). The Zimran score showed a moderate negative correlation with the following domains of the SF-36: physical functioning ($p = 0.035$), role-physical ($p = 0.036$), general health ($p = 0.023$) and role emotional ($p = 0.021$).

Discussion and Conclusion: Although limited because of the small number of patients included, findings suggest that patients with GD receiving ERT have a better QoL than patients with FD or with GD not receiving ERT. Imiglucerase has a beneficial effect against pain for patients with GD. Further studies should be conducted to confirm our findings.

Communicated by: Ed Wraith

Competing interests: None declared

F.L. Oliveira · B.C. Krug · I.V.D. Schwartz
Post-Graduate Program in Medical Sciences, School of Medicine,
Universidade Federal do Rio Grande do Sul (UFRGS), Porto
Alegre, Brazil

T. Alegra
Post-Graduate Program in Genetics and Molecular Biology,
Universidade Federal do Rio Grande do Sul (UFRGS), Porto
Alegre, Brazil

A. Dornelles · C.B.O. Netto · I.V.D. Schwartz (✉)
Medical Genetics Service, Hospital de Clínicas de Porto Alegre
(HCPA), Rua Ramiro Barcellos, 2350,
Porto Alegre, RS 90035-003, Brazil
e-mail: ischwartz@hcpa.ufrgs.br

I.V.D. Schwartz
Department of Genetics, UFRGS, Porto Alegre, Brazil

N.S. da Rocha
Department of Psychiatry and Legal Medicine, UFRGS, Porto
Alegre, Brazil

P.D. Picon
Department of Internal Medicine, HCPA, State Health Department
of Rio Grande do Sul, Porto Alegre, Brazil

Introduction

The concept of quality of life (QoL) as a health outcome measure was introduced in the 1970s and has evolved since then (Panzini and Bandeira 2005). According to the World Health Organization (WHO) (1994), quality of life is an individual's perception of their position in life in the context of the culture and value system in which they live and in relation to their goals, expectations, standards, and concerns. According to this definition, disease may affect an individual's health as well as other general aspects of their life.

Lysosomal storage diseases (LSD) are rare inherited disorders caused by specific enzyme deficiencies that lead to an abnormal storage of normal substrates or their catabolic products in the lysosomes (Meikle et al. 1999a). LSDs form a group of about 50 diseases (Wraith 2002) with an estimated incidence of 1:7,000 live births (Meikle et al.

1999b; Poorthuis et al. 1999). LSDs are classified according to the substrate stored: sphingolipidoses, such as Gaucher disease (GD) or Fabry disease (FD); mucopolysaccharidoses (MPS); glycoproteinoses; and others (Gieselmann 1995; Raas-Rothschild et al. 2004). Enzyme administration of the recombinant form of the enzyme that is deficient in LSD patients is known as enzyme replacement therapy (ERT). ERT is currently available for the following LSDs: GD, FD, MPS I, MPS II, MPS VI, and Pompe disease. The effects of ERT on the QoL of patients with GD have already been described (Damiano et al. 1998; Masek et al. 1999; Giraldo et al. 2000; Pastores et al. 2003; Giraldo et al. 2005; Weinreb et al. 2007), but are still unknown for patients with FD.

No studies in the literature have evaluated the QoL of Brazilian patients with LSDs, and no disease-specific instrument to evaluate their QoL has been evaluated, translated, or validated.

This study used the SF-36 survey to evaluate the QoL of patients with GD or FD seen in the Medical Genetics Service of Hospital de Clínicas de Porto Alegre, Porto Alegre, Brazil (SGM-HCPA), a national reference center for the diagnosis and treatment of LSD located in southern Brazil.

Methods

All patients with GD or FD seen at the SGM-HCPA were invited to participate in this observational cross-sectional study during their routine follow-up visits from September to October 2008. Patients should meet all of the following inclusion criteria: (a) age 12 years or older; (b) cognitive ability to fill out the survey; (c) signature of an informed consent form (ICF) agreeing to participate in the study. In the case of patients younger than 18 years, the consent form should also be signed by a parent or guardian.

Patients were evaluated using the SF-36 survey, which was administered after an interview to collect general data, such as type of LSD, current age, age at diagnosis, and time on ERT. Clinical severity of GD was defined using the Zimran (Zimran et al. 1992) severity score index recorded in the patient's chart for each patient on the date closest to the time the SF-36 was administered. The Zimran score is calculated according to the following disease characteristics: presence of cytopenia; hepatosplenomegaly; splenectomy; involvement of the central nervous system; skeletal findings defined by radiological, scintigraphic and clinical findings such as the occurrence of pain; involvement of other organs such as the lungs; as well as liver disease. Higher Zimran scores indicate greater clinical severity (mild $= 0$–10; moderate $= 11$–19; severe $= \geq20$). In the case of FD, the medical team at SGM-HCPA at the time of this study did not use severity scores when evaluating patients.

Medical Outcomes Study: 36-Item Short Form Health Survey (SF-36)

SF-36 is a generic instrument (Ware and Sherbourne 1992) that has 36 items organized in eight domains. Four of these domains provide a physical component score: physical functioning, which assesses the presence and severity of limitations associated with physical capacities; role limitations due to physical health, which assesses limitations according to type and amount of work, as well as how much these limitations affect work and activities of daily living; bodily pain, which assesses the presence of pain, its intensity, and how it affects activities of daily living; and general health, which assesses how patients feel about their personal health in general. The other four provide a mental component score: role limitations due to emotional problems, which assesses the impact of psychological aspects on the patient's well-being; vitality, which takes into consideration the level of energy and fatigue; social functioning, which analyzes the integration of the individual in social activities; and mental health, which includes questions about anxiety, depression, changes in behavior or emotional unbalance, and psychological well-being (Ware et al. 1993; McHorney et al. 1993; Hsiung et al. 2005). The survey has dichotomous or ordinal response items and should be answered considering a 4-week period before its administration (Ware et al. 1993; Weinreb et al. 2007). Higher scores indicate better QoL.

Statistical Analysis

Means and standard deviations were used to describe the quantitative variables, which were analyzed for the whole group of patients in the study, as well as for sample subgroups. The Kolmogorov-Smirnov test was used to check whether data distribution was significantly different from normal. Because most SF-36 items showed normal distribution, and all studies published so far on SF-36 in GD have used means to describe them, we chose to use means in all cases.

No studies in the literature reviewed evaluated QoL using the SF-36 in patients with FD receiving ERT. For patients with GD receiving ERT, six studies were retrieved (Damiano et al. 1998; Masek et al. 1999; Giraldo et al. 2000, 2005; Pastores et al. 2003; Weinreb et al. 2007). The study conducted by Masek et al. (1999) with North American patients with GD was used for comparisons with our study. The Student t test was used to compare QoL of Brazilian and American patients with GD receiving ERT. For the comparison with a normal population, the reference was the group of elderly patients studied by Lima et al. (2009) because similar studies with Brazilian adults defined as healthy were not found in the literature.

Table 1 Characteristics of the sample included in the present study ($n = 35$)

Characteristic	Gaucher disease $n = 21$ (type I = 20; III = 1)	Fabry disease $n = 14$
Age (mean ± SD) years	24.9 ± 13.4	37.3 ± 12.9
Gender (female/male)	11/10	4/10
Time on ERT (mean ± SD) years	8.5 ± 4.5	3.5 ± 2.1
n	15 (imiglucerase)	10 (agalsidase alpha = 7; agalsidase beta = 3)

SD Standard deviation

The level of statistical significance was set at 5% for all analyses. Statistical calculations were made using the software SPSS® for Windows® 18.0.

Definition of Clinically Significant Change

Changes in domain scores were classified as clinically significant according to the study conducted by Kosinski et al. (2000), who evaluated patients with rheumatoid arthritis in the USA. The minimally significant changes defined by those authors were as follows: physical functioning – 8.4; role-physical – 21.0; bodily pain – 14.7; general health – 4.2; vitality – 11.1; social functioning – 11.7; role-emotional – 17.9; and mental health domain – 7.3. The same criteria were used by Weinreb et al. (2007) in their study about the evaluation of QoL in patients with GD.

Results

Thirty-five patients were included in the study. Only one patient with GD type I did not agree to participate in the study due to personal reasons. Patient characteristics are summarized in Table 1. Of the patients included, only six, all with GD, were younger than 18 years (4/6 were 12–14 years).

Gaucher Disease ($n = 21$)

Mean Zimran score was 6.02 (range = 1–29); the disease was mild in 19 patients, moderate in 1, and severe in 1. There was a statistically significant correlation between the Zimran score and the following SF-36 domains: physical functioning ($r = -0.462$; $p = 0.035$); role-physical ($r = -0.460$; $p = 0.036$), general health ($r = -0.494$; $p = 0.023$), and role-emotional ($r = -0.501$; $p = 0.021$).

In the group of patients receiving ERT ($n = 15$), the mean SF-36 domain scores ranged from 67.6 (vitality) to 77.3 (mental health); for the patients not receiving ERT ($n = 6$), from 38.8 (role-emotional) to 72.0 (mental health) (Tables 2 and 3). The comparison of patients on ERT and patients naïve to ERT revealed that bodily pain was the only domain whose score had a statistically significant difference between groups ($p = 0.036$), and the higher scores were found to be for the ERT group. According to the criteria established by Kosinski et al. (2000), this change may also be classified as clinically significant. The physical functioning, role-physical, general health, social functioning, and role-emotional domains also showed clinically significant changes that favored the ERT subgroup (Table 3).

Fabry Disease ($n = 14$)

In the group of patients with FD, eight were receiving ERT with agalsidase alpha and four with agalsidase beta (Table 1). In the analysis of male patients only ($n = 10$, all receiving ERT), mean domain scores ranged from 45.0 (role-physical) to 66.6 (role-emotional) (Table 3). A summary of clinical manifestations and domain scores for heterozygous women included in the study is given in Table 4.

Comparisons Between GD and FD Patients Receiving ERT and Other Populations

The comparison of patients with GD and FD receiving ERT showed a statistically significant difference in the general health domain ($p = 0.046$), which was higher in patients with GD. The comparison with other GD patients and with a Brazilian normal population is shown in Tables 2 and 3.

Discussion

This is the first study to evaluate QoL of Brazilian patients with LSD using the SF-36 survey. Our findings suggest that the QoL of patients with GD and FD is significantly affected and that ERT has a beneficial effect on the QoL of patients with GD.

The SF-36 survey is, in fact, a general instrument to evaluate health-related quality of life that has not been validated for use in patients with LSD. However, it was used in our study for two main reasons: first, most studies in the literature published about this issue until 2008 used the SF-36 survey; second, this questionnaire has been validated for the Brazilian population (Ciconelli et al. 1997). These reasons

Table 2 Comparison of quality of life of Brazilian and North American patients with GD receiving enzyme replacement therapy, according to SF-36 survey responses[*]

	Brazilian patients with Gaucher disease ($n = 15$)	American patients with Gaucher disease (Masek et al. 2009) ($n = 25$)	**p**
Mean age (years)	22.3	41.7	
Mean treatment time (years)	8.5	2	
SF-36			
Physical component			
Physical functioning	76.3 ± 26.0	76.7 ± 29.5	0.965
Role-physical	70.0 ± 42.4	80.4 ± 32.8	0.318
Bodily pain	76.5 ± 27.3	66.3 ± 25.6	0.241
General health	71.1 ± 21.6[**]	59.3 ± 24.0[**]	0.126
Mental component			
Vitality	67.6 ± 20.6[**]	56.3 ± 23.3[**]	0.129
Social functioning	70.0 ± 20.9[**]	87.0 ± 18.6[**]	0.011[a]
Role-emotional	68.8 ± 42.6	75.3 ± 36.6	0.612
Mental health	77.3 ± 9.7	73.9 ± 14.5	0.426

[*] Data described as means and standard deviations for the SF-36 domains.

[**] Clinically significant changes for the SF-36 scores defined according to the criteria for patients with rheumatoid arthritis in the USA (see Methods)

[a] $p < 0.05$

Table 3 Comparison of quality of life of Brazilian patients with Gaucher and Fabry diseases and a group of Brazilian elderly according to the SF-36[a]

	Gaucher disease receiving ERT ($n = 15$)	Gaucher disease not receiving ERT ($n = 6$)	Fabry disease receiving ERT ($n = 10$)	Elderly(Lima et al. 2005) ($n = 1.958$)[b]
Mean age (years)	22.3 ±13.0	31.5 ±13.1	33.5 ± 12.9	60 ± 69.6
SF-36				
Physical component				
Physical functioning	76.3 ± 26.0[b]	59.1 ± 38.3[a,b]	66.0 ± 26.5	71.4
Role-physical	70.0 ± 42.4[b]	41.6 ± 46.5[a,b]	45.0 ± 44.7[a]	81.2
Bodily pain	76.5 ± 27.3[b,c]	47.8 ± 22.7[a,b,c]	59.7 ± 28.5	74.2
General health	71.1 ± 21.6[b,d]	62.5 ± 19.1[a,b]	50.3 ± 26.8[a,d]	70.1
Mental component				
Vitality	67.6 ± 20.6[b]	56.6 ± 16.3[b]	63.5 ± 21.6	64.4
Social functioning	70.0 ± 20.9[a,b]	55.2 ± 31.2[a,b]	65.0 ± 25.4[a]	86.1
Role-emotional	68.8 ± 42.6[b]	38.8 ± 49.0[a,b]	66.6 ± 41.3[a]	85.9
Mental health	77.3 ± 9.7[a]	72.0 ± 11.8	64.4 ± 25.3	69.9

[*] Data described as means and standard deviations for SF-36 components and domains

[**] Standard deviations of SF-36 scores not reported in original study

[a] Clinically significant differences in comparison with group of elderly patients

[b] Clinically significant differences between the groups of patients with GD receiving or not receiving ERT

[c] $p < 0.05$ (GD receiving ERT x GD not receiving ERT)

[d] $p < 0.05$ (GD receiving ERT × FD receiving ERT)

should be taken into consideration when discussing our findings because general instruments, such as the SF-36 survey, may not be sensitive enough to detect specific patterns of QoL impairments. In addition, although the SF-36 survey has been validated for individuals older than 18 years, the Gaucher Registry suggests its administration for individuals as young as 14 years, and there is at least one study about this survey used for patients with GD 14 years and older (Damiano et al. 1998). Our study included individuals at least 12 years old, but most were 14 years or

Table 4 Fabry disease – summary of clinical characteristics and mean SF-36 scores of heterozygous women included in the study

Patient	Age (years)	Clinical manifestations	ERT	PF	RP	BP	GH	VT	SF	RE	MH
A	33	Acroparesthesia, angiokeratoma, verticillate cornea, proteinuria	No	100	100	72	100	75	100	66	84
B	41	Acroparesthesia, angiokeratoma, depression, migraine, proteinuria, hematuria	No*	75	100	74	72	70	87.5	100	80
C	52	Acroparesthesia, depression, joint pain	Yes	20	0	31	25	35	84.3	0	40
D	61	Hypertension, transient ischemic attack, tinnitus, dizziness, intolerance to heat and cold	Yes	90	100	100	52	85	87.5	100	92

ERT enzyme replacement therapy, *PF* physical functioning, *RP* role-physical, *GH* general health, *VT* vitality, *SF* social functioning, *RE* role-emotional, *MH* mental health

*Patient started ERT immediately after responding to SF-36

older. The lower limit of age adopted as an inclusion criteria was justified by the need to have a larger study sample and because individuals 12 years and older are usually capable of responding questionnaires such as the SF-36 (Streiner et al. 1995).

Our sample size made it impossible to analyze the effects of ERT on FD or the analysis of a possible differential effect between the two forms of agalsidase.

The lack of similar studies for comparisons has also limited our analysis. The comparison with a sample of Brazilian elderly individuals (Lima et al. 2009) revealed that patients with GD receiving ERT had clinically significant differences in social functioning, with a higher score for the elderly, and mental health, with a higher score for the patients with GD. The patients with GD not receiving ERT had clinically significant differences in all the domains, (with higher scores for the elderly), except in the vitality and mental health domains. In the group of patients with FD receiving ERT, clinically significant differences were found in the role-physical, general health, social functioning, and role-emotional domains, all with higher scores for the elderly.

Gaucher Disease

According to our results, ERT has a significant and positive effect on the QoL of GD patients. These findings suggest that the generic approach to evaluate QoL was sensitive to detect changes in the health of patients with GD on ERT or ERT-naïve, and corroborate findings by Masek et al. (1999), who compared the QoL of the general population in the USA with that of patients with GD receiving ERT. Their results demonstrated a significant improvement in vitality, role-physical, and social functioning. Giraldo et al. (2005) evaluated the QoL of patients after 2 years of treatment and the importance of bone pain. They showed that role-physical and bodily pain are negatively associated with QoL, but failed to demonstrate that ERT improved QoL for their patients. Weinreb et al. (2007) evaluated ERT efficacy before

and after 4 years of treatment. They reported that treatment had a positive effect on the QoL of patients with GD.

Patients with GD not receiving ERT had lower scores in the role-emotional domain of the mental component score; in the physical component score, the lowest score was in the role-physical domain. These domains are associated with depression, anxiety, and poor physical performance reported by patients when not able to complete or keep the pace of their daily activities, such as their work activities (McHorney et al. 1993). The domain with the highest score was mental health, also in the group receiving ERT. This finding may be explained by the fact that all patients included in this study, except one, had GD type I (the non-neuropathic form).

The comparison between Brazilian and North American patients with GD receiving ERT (Masek et al. 1999) revealed clinically significant changes in general health and vitality (higher scores in the Brazilian group) and social functioning (higher score in the American group). The better findings in the Brazilian patients may be partially explained by the differences in age and treatment duration between the two populations, as the Brazilians had a lower mean age and a longer time on ERT. The scores found in social functioning might be explained by cultural (or socioeconomic) differences between the two samples. Socioeconomic data or mean imiglucerase dose received by the patients could not be compared because Masek et al. (1999) did not report those findings.

The analysis of Zimran scores revealed a significant negative, as expected, but moderate correlation, particularly with the physical domain. The only physical domain that had no significant correlation was bodily pain, exactly the one that, in our study, showed a significant positive difference in the group receiving ERT. According to the definition of relevant clinical change adopted in this study, patients with GD receiving or not receiving ERT also had a difference, although not statistically significant, in the three domains of the physical component that had a significant correlation with the Zimran score (physical functioning, role-physical, general health) and in two domains of the mental component of the SF-36 (social functioning and role-emotional), but only the role-emotional domain had a significant correlation with the Zimran score. These findings suggest that the Zimran score, originally developed to evaluate the severity of disease in patients with GD in studies about the association between genotype and phenotype, is an adequate instrument to detect mainly physical changes associated with GD, but not to detect the effect of the treatment.

Fabry Disease

Unlike GD, FD is an X-linked disorder, and heterozygous women may be either asymptomatic, oligosymptomatic, or have a disease similar to that found in hemizygotes (Pinto et al. 2010). As expected, the heterozygous women in this study usually had higher SF-36 scores than men.

Male patients receiving ERT had the lowest scores in the role-physical domain of the physical component score. This domain is associated with the number of times that the patient had to reduce daily activities due to physical problems. Patient activities may be impaired due to associated clinical comorbidities, such as depression, paresthesias, or heart disease. In the mental component score, the lowest result was in the vitality domain, which is associated with the levels of energy and fatigue (McHorney et al. 1993); this finding might also be associated with the psychiatric abnormalities seen in the patients.

Conclusion

Data in this study suggest that patients with GD receiving ERT have a better QoL than patients with FD and GD not receiving ERT and that ERT with imiglucerase has a beneficial effect against the pain felt by patients. In the group of patients with FD, scores were higher in the group of heterozygous women, which may be explained by the fact that FD is an X-linked disorder. Further studies should be conducted to confirm our findings.

Acknowledgments The authors thank the staff of the Medical Genetics Service of HCPA and the staff of the Reference Center for Gaucher Disease for their support and collaboration in the conduction of this study. We also thank the Research and Graduate Program of HCPA, Porto Alegre, Brazil, for their support and preparation of statistical analysis, especially the statistician Marilyn Agranonik.

Synopsis of the Article

This study evaluated the quality of life of patients with Gaucher and Fabry diseases using a nonspecific instrument, the SF-36 questionnaire. Results suggested that ERT has a positive effect against the pain domain for patients with Gaucher disease.

Authors Contributions

Fabiane Lopes Oliveira Conception and design, analysis and interpretation of data, drafting of the article.

Taciane Alegra Analysis and interpretation of data, drafting of the article.

Alicia Dornelles Analysis and interpretation of data, drafting of the article.

Bárbara Corrêa Krug Analysis and interpretation of data, drafting of the article.

Cristina B. O. Netto Analysis and interpretation of data, drafting of the article.

Neusa Sica da Rocha Analysis and interpretation of data, drafting of the article.

Paulo D. Picon Analysis and interpretation of data, drafting of the article.

Ida Vanessa D. Schwartz Conception and design, analysis and interpretation of data, drafting of the article.

Guarantor for the Article

Ida Vanessa D. Schwartz.

Competing Interest

All authors declare that the answer to all questions on the JIMD competing interest form is No, and therefore they have nothing to declare.

Funding

This study was partially funded by FIPE/HCPA, Porto Alegre, Brazil (project # 08-209), and by grants from the Brazilian Agency of Research-CNPq (MCT/CNPq/ MS-SCTIE-DECIT 33/2007; MCT/CNPq/MS-SCTIE-DECIT 37/2008; and MCT/CNPq/MS-SCTIE-DECIT 67/2009).

Ethics Approval

This study was approved by the Ethics in Research Committee of the Hospital de Clínicas de Porto Alegre, Porto Alegre, Brazil.

Patient Consent

All patients or their guardians signed an informed consent form.

References

Ciconelli RM, Ferraz MB, Santos W, Meinão I, Quaresma MR (1997) Brazilian-Portuguese version of the SF-36. A reliable and valid quality of life outcome measure. Rev Bras Reumatol 39:50

Damiano AM, Pastores GM, Ware JE (1998) The health-related quality of life of adults with Gaucher's disease receiving enzyme replacement therapy: results from a retrospective study. Qual Life Res 7:373–386

Gieselmann V (1995) Lysosomal storage diseases. Biochim Biophys Acta 1270:103–136

Giraldo P, Pocoví M, Pérez-Calvo J, Rubio-Félix D, Giralt M (2000) Report of the Spanish Gaucher's Disease Registry: clinical and genetic characteristics. Haematologica 85:792–799

Giraldo P, Solano V, Pérez-Calvo JI, Giralt M, Rubio-Félix D, Spanish Group on Gaucher Disease (2005) Quality of life related to type 1 Gaucher disease: Spanish experience. Qual Life Res 14:453–462

Hsiung PC, Fang CT, Chang YY, Chen MY, Wang JD (2005) Comparison of WHOQOL-bREF and SF-36 in patients with HIV infection. Qual Life Res 14:141–150

Kosinski M, Zhao SZ, Dedhiya S, Osterhaus JT, Ware JE Jr (2000) Determining minimally important changes in generic and disease-specific health-related quality of life questionnaires in clinical trials of rheumatoid arthritis. Arthritis Rheum 43:1478–1487

Lima MG, Barros MB, Cesar CL, Goldbaum M, Carandina L, Ciconelli RM (2009) Health related quality of life among the elderly: a population-based study using SF-36 survey. Cad Saude Publica 25:2159–2167

Masek BJ, Sims KB, Bove CM, Korson MS, Short P, Norman DK (1999) Quality of life assessment in adults with type 1 Gaucher disease. Qual Life Res 8:263–268

McHorney CA, Ware JE Jr, Raczek AE (1993) The MOS 36-Item Short-Form Health Survey (SF-36): II. Psychometric and clinical tests of validity in measuring physical and mental health constructs. Med Care 31:247–263

Meikle PJ, Hopwood JJ, Clague AE, Carey WF (1999a) Prevalence of lysosomal storage disorders. JAMA 281:249–254

Meikle PJ, Ranieri E, Ravenscroft EM, Hua CT, Brooks DA, Hopwood JJ (1999b) Newborn screening for lysosomal storage disorders. Southeast Asian J Trop Med Public Health 30(Suppl 2):104–110

Panzini RG, Bandeira DR (2005) Quality of life and spiritual-religious coping relations. Qual Life Res 14:2106–2107

Pastores GM, Barnett NL, Bathan P, Kolodny EH (2003) A neurological symptom survey of patients with type I Gaucher disease. J Inherit Metab Dis 26:641–645

Pinto LL, Vieira TA, Giugliani R, Schwartz IV (2010) Expression of the disease on female carriers of X-linked lysosomal disorders: a brief review. Orphanet J Rare Dis 5:14

Poorthuis BJ, Wevers RA, Kleijer WJ et al (1999) The frequency of lysosomal storage diseases in The Netherlands. Hum Genet 105:151–156

Raas-Rothschild A, Pankova-Kholmyansky I, Kacher Y, Futerman AH (2004) Glycosphingolipidoses: beyond the enzymatic defect. Glycoconj J 21:295–304

Streiner DL, Norman GR (1995) Health measurement scales: a practical guide to their development and use. Oxford University Press, Oxford

The WHOQOL Group (1994) Development of the WHOQOL: rationale and current status. Int J Ment Health 23:24–56

Ware JE Jr, Sherbourne CD (1992) The MOS 36-item short-form health survey (SF-36). I. Conceptual framework and item selection. Med Care 30:473–483

Ware JE, Snow KK, Kosinski M et al (1993) SF-36 health survey. Manual and interpretation guide. New England Medical Center, Boston

Weinreb N, Barranger J, Packman S et al (2007) Imiglucerase (Cerezyme) improves quality of life in patients with skeletal manifestations of Gaucher disease. Clin Genet 71:576–588

Wraith JE (2002) Lysosomal disorders. Semin Neonatol 7:75–83

Zimran A, Kay A, Gelbart T, Garver P, Thurston D, Saven A, Beutler E (1992) Gaucher disease. Clinical, laboratory, radiologic, and genetic features of 53 patients. Medicine 71:337–53

JIMD Reports
DOI 10.1007/8904_2012_138

Identification and Functional Characterization of *GAA* Mutations in Colombian Patients Affected by Pompe Disease

Mónica Yasmín Niño · Heidi Eliana Mateus · Dora Janeth Fonseca ·
Marian A. Kroos · Sandra Yaneth Ospina · Juan Fernando Mejía ·
Jesús Alfredo Uribe · Arnold J. J. Reuser · Paul Laissue

Received: 27 January 2012 / Revised: 27 January 2012 / Accepted: 24 February 2012 / Published online: 19 April 2012
© SSIEM and Springer-Verlag Berlin Heidelberg 2012

Abstract Pompe disease (PD) is a recessive metabolic disorder characterized by acid α-glucosidase (GAA) deficiency, which results in lysosomal accumulation of glycogen in all tissues, especially in skeletal muscles. PD clinical course is mainly determined by the nature of the *GAA* mutations. Although ~400 distinct *GAA* sequence variations have been described, the genotype-phenotype correlation is not always evident.

In this study, we describe the first clinical and genetic analysis of Colombian PD patients performed in 11 affected individuals. *GAA* open reading frame sequencing revealed eight distinct mutations related to PD etiology including two novel missense mutations, c.1106 T > C (p.Leu369Pro) and c.2236 T > C (p.Trp746Arg). In vitro functional studies showed that the structural changes conferred by both mutations did not inhibit the synthesis of the 110 kD GAA precursor form but affected the processing and intracellular transport of GAA. In addition, analysis of previously described variants located at this position (p.Trp746Gly, p.Trp746Cys, p.Trp746Ser, p.Trp746X) revealed new insights in the molecular basis of PD. Notably, we found that p.Trp746Cys mutation, which was previously described as a polymorphism as well as a causal mutation, displayed a mild deleterious effect. Interestingly and by chance, our study argues in favor of a remarkable Afro-American and European ancestry of the Colombian population. Taken together, our report provides valuable information on the PD genotype–phenotype correlation, which is expected to facilitate and improve genetic counseling of affected individuals and their families.

Communicated by: Robin Lachmann

Competing interests: None declared

Authors MónicaYasmín Niño and Heidi Eliana Mateus contributed equally to this work.

M.Y. Niño · H.E. Mateus · D.J. Fonseca · S.Y. Ospina · P. Laissue
Unidad de Genética, Escuela de Medicina y Ciencias de la Salud,
Universidad del Rosario, Bogotá, Colombia

M.Y. Niño · M.A. Kroos · A.J.J. Reuser
Department of Clinical Genetics, Center for Lysosomal and
Metabolic Diseases, Erasmus MC, Rotterdam, The Netherlands

J.F. Mejía
Fundación Valle de Lili, Cali, Colombia

J.A. Uribe
Departamento de Ciencias Biológicas, Universidad de los Andes,
Bogotá, Colombia

A.J.J. Reuser (✉)
Department of Clinical Genetics, Erasmus MC University
Medical Center, Dr Molewaterplein 50,
3015 GE, Rotterdam, The Netherlands
e-mail: a.reuser@erasmusmc.nl

Introduction

Pompe disease (PD) is a rare autosomal recessive metabolic disorder characterized by acid α-glucosidase (GAA) deficiency. This enzyme catalyzes the hydrolysis of the α-1,4 and α-1,6-glucosidic bonds of glycogen, and its deficiency results in lysosomal glycogen storage in all tissues, especially in skeletal muscles. Clinically, PD patients display a broad spectrum of phenotypes with regard to the age of onset, the disease progression rate, and the severity of symptoms (van der Ploeg and Reuser 2008; Raben et al. 2007; Hirschhorn et al. 2001). The infantile form of PD includes severely affected infants (under 1 year of age) who display a combination of generalized skeletal muscle weakness and cardiac hypertrophy that provoke cardiorespiratory failure and death (Kishnani et al. 2006; van den Hout 2003). Conversely, patients with childhood, juvenile, and adult onset of symptoms lack cardiac involvement. These individuals exhibit a less severe skeletal muscle dysfunction with slowly progressive proximal myopathy as well as a marked involvement of respiratory muscles (Hirschhorn et al. 2001; van der Ploeg 2008; Müller-Felber et al. 2007; Laforêt et al. 2000). The *GAA* gene contains 19 coding exons. At the protein level, human GAA is composed of five distinct regions: trefoil type-P, N-terminal β-sandwich, catalytic $(\beta/\alpha)_8$ barrel, proximal C-terminal, and distal C-terminal domains. The key catalytic residues are located at Asp518 and Asp616 (Sugawara et al. 2009). GAA is synthesized as an inactive precursor of 110 kD which is subsequently transported to the pre-lysosomal and lysosomal compartments via the mannose 6-phosphate receptor. En route, it is processed to a 95 kD intermediate and subsequently to fully active forms of 76 and 70 kD (Hirschhorn et al. 2001).

From an etiological point of view, PD is caused by *GAA* mutations that determine the degree of enzyme deficiency and largely the clinical course (Kroos et al. 2008; Reuser et al. 1987). Disease-causing sequence variations have been described along the entire length of the gene including missense and nonsense mutations, splice site variants, and partial insertions/deletions. At present, the Pompe Disease Mutation database (www.pompecenter.nl) lists 393 *GAA* sequence variations. Among these, 54 are of unknown effect, 75 are considered as nonpathogenic, 2 are probably non-pathogenic, and 257 are confirmed as etiological (Oba-Shinjo et al. 2009; http://www.pompecenter.nl/). Several in vitro studies have permitted to propose distinct molecular mechanisms underlying PD etiopathology (Kroos et al. 2008). The majority of pathogenic missense mutations seem to affect folding, posttranslational processing, and/or intracellular transport of GAA which partially or completely abolishes its function (Pittis et al. 2008; van der Ploeg and Reuser 2008).

Some *GAA* mutations seem to have spread through a founder effect. African American patients originating from the north of Africa frequently present c.2560 C > T (p.Arg854X) and Asian patients c.1935 C > A (p.Asp645Glu) sequence variants (Becker et al. 1998). Common mutations among Caucasian patients include c.2481 + 102_2646del (delexon18; p.Gly828_Asn882del), c.525del (delT525; p.Glu176fsX45), and c.925 G > A (p.Gly309Arg) (Hirschhorn et al. 2001; Kroos et al. 2008; Raben et al. 1999). The c.-32-13 T > G mutation, which reduces the *GAA*-mRNA splicing fidelity, is the most common *GAA* pathogenic sequence variant among Caucasian adults and children with a slowly progressive course of the disease (Boerkoel et al. 1995; Huie et al. 1994). The GAA residual activity in patients presenting the c.-32-13 T > G/null genotype is usually reduced to 5–25% of average normal (van der Ploeg and Reuser 2008; Kroos et al. 2007). Some patients with this genotype manifest symptoms in early childhood, whereas others remain presymptomatic until late adulthood. This demonstrates the role of modifying factors in PD pathophysiology (Pittis et al. 2008; Kroos et al. 2007; Slonim et al. 2007).

In this study, we describe the first clinical and genetic analysis of Colombian PD patients performed in 11 affected individuals who belong to 8 families. Direct sequencing of the complete *GAA* open reading frame revealed eight distinct mutations related to PD etiology. Two novel missense mutations were investigated for their functional effect along with four previously described mutations to obtain a better understanding of the disease pathophysiology. Interestingly and by chance, our study argues in favor of a remarkable Afro-American and European ancestry of the Colombian population.

Material and Methods

Patients

PD patients (pt) who belong to eight distinct families were included in this study. These individuals originate from five different Colombian cities: Cartagena, Barranquilla, Bucaramanga, Medellín, and Bogotá. As previously described, PD diagnosis was performed by quantifying GAA activity from peripheral blood leukocytes using 4-methylumbelliferyl-α-D-glucoside as substrate (Li et al. 1785). Maltase-glucoamylase activity was inhibited with 120 μmol/L of acarbose. For each patient, GAA activity was assayed in the presence and absence of acarbose at pH 3.8. More than 85% inhibition confirms PD (Palmer et al. 2007). The age at diagnosis ranged from 2 to 47 years and the initial symptoms were mostly related to limb girdle weakness (LW) (Table 1). Ten patients (pt 1 to pt 10) experienced first symptoms in childhood or adulthood. One patient (pt 11) was diagnosed at 4 months of age since he displayed hypotonia and delayed motor development. Pt 4, 5, and 6 belong to one family. The same holds for patients 9 and 10. In all the cases, parents of the patients were included in the study in order to evaluate the segregation of the *GAA* mutations. All participants in the study provided written informed consent. The Institutional Ethics Committee of each participating institution approved the clinical and experimental aspects of the study.

GAA Mutational Analysis

Genomic DNA was extracted from whole blood samples using standard procedures. In all patients, the complete *GAA* open reading frame (19 exons) was amplified by PCR as previously described (Becker et al. 1998). Each amplicon

Table 1 Phenotypes and Genotypes of Colombian Pompe Disease Patients

	Patient 1	Patient 2	Patient 3	Patient 4	Patient 5	Patient 6	Patient 7	Patient 8	Patient 9	Patient 10	Patient 11
Clinical features											
Gender	M	M	F	F	M	M	M	F	M	M	M
First symptoms (years)	1.1	6	12	20	19	17	27	38	10	15	0.3
Age at diagnosis (years)	5	11	29	33	35	36	31	47	19	20	2
Initial clinical signs	PI	LW	LW	LW	LW	LW	LW	LW	LW	LW	HDMD
Muscle weakness	LL	WC	LL	LL	LL	LL	LL	LL	LL	WC	RPMW
Respiratory distress	VS	VS	No	No	No	No	O	O and FE	VS/night	VS	VS
Cardiomyopathy	No	No	No	No	No	No	No	No	No	No	Yes
Hepatomegaly	Yes	No	No	No	No	No	No	No	No	No	No
Underweight	Yes	No	No	No	No	No	No	No	No	No	Yes
Associated	FD	SDB	No	No	No data	No data	No	No	SDB	FD	FD
CK (IU/L)	127	545	332	No data	No data	No data	488	912	977	736	350
AST/ALT (IU/L)	142/144	187/112	66/70	No data	No data	No data	56/46	46/44	125/118	213/154	226/180
GAA mutations	c.1064T>C (p.Leu355 Pro) c.1106T>C (p.Leu369 Pro)*	c.1064T>C (p.Leu355 Pro)	c.-32-13T>G c.525delT (p.Glu176 fsx45)	c.2481+102_2646del (p.Gly828_Asn882del) c.-32-2A>G	Idem Pt4 Idem Pt4	Idem Pt4 Idem Pt4	c.2560 C>T (p.Arg854X) c.1581 A>G (p.Arg527Arg) c.-32-13T>G c.596G >A (p.Arg199His) c.668A >G (p.His223 Arg)	c.-32-13T>G	c.2560 C>T (p.Arg 854X) c.1551 +42A>G c.1581A >G (p.Arg527 Arg) c.596G >A (p.Arg199 His) c.668A >G (p.His223 Arg)	c.2560 C>T (p.Arg 854X) c.1551 +42A>G c.1551 +42A>G c.596G >A (p.Arg199 His) c.668A >G (p.His223 Arg)	c.2560C>T (p.Arg854X) c.2236 T>C (p.Trp746 Arg)* c.596G >A (p.Arg 199His) c.68A >G (p.His223 Arg) c.2553A >G (p.Gly851 Gly)

(continued)

Table 1 (continued)

	Patient 1	Patient 2	Patient 3	Patient 4	Patient 5	Patient 6	Patient 7	Patient 8	Patient 9	Patient 10	Patient 11
							c.1551 +49A>C		c.2338A >G (p.Ile780 Val)	c.2338A >G (p.Ile780 Val)	c.1551+49A>C
							c.2338A >G (p. Ile780Val)		c.2553A >G(p. Gly851 Gly)	c.2553A >G (p. Gly851 Gly)	
Laboratory findings											
CK (IU/L)	127	545	332	No data	No data	No data	488	912	977	736	350
AST/ALT (IU/L)	142/144	187/112	66/70	No data	No data	No data	56/46	46/44	125/118	213/154	226/180

PI Pulmonary, *LW* Limbgirdle weakness, *HDMD* Hypotonia/delayed motor development, *LL* Lower limbs, *WC* Wheelchair; *O* Orthopnea, *FE* Fatigue on exertion, *FD* Feeding difficulties, *SDB* Sleep-disordered breathing, *RPMW* Rapidly progressive muscle weakness.

Underlined mutations denote homozygous status; mutations in bold are pathogenic; *denotes novel mutations.

was purified by using shrimp alkaline phosphatase and exonuclease I. PCR primers were used to sequence the coding regions in both sense and antisense directions using an ABI 3730xl sequencer (Ko et al. 1999). The presence of each non-synonymous variant was confirmed by an additional round of PCR and sequencing. Variations at the DNA level were identified using human *GAA* wild-type mRNA sequence (NM_000152.3). Sequence variations were described according to the international mutation nomenclature guidelines as set forth by the Human Genome Variation Society (http://www.hgvs.org/mutnomen/). Intron mutations were designated by locating its cDNA position and, as described by den Dunnen et al. (den Dunnen and Antonarakis 2000), negative numbers were reported from the starting of the splice acceptor site. GAA mutant protein sequences were aligned and compared with the human wild-type version (NP_000143.2) using ClustalW software. In order to assess conservation during evolution of residues at mutated sites, this program was also used to perform multiple alignments of protein sequences from vertebrate species: *Homo sapiens*, *Pongo abeili*, *Bos taurus*, *Mus musculus*, and *Rattus norvegicus*. Data from the Pompe disease mutation database (www.pompecenter.nl) were used to define novel *GAA* sequence variants as well as to identify which of them were previously related to pathogenic effects. To predict the effect of newly identified missense mutations we also used SIFT and PolyPhen2 software. PolyPhen2 prediction values are the result of an algorithm, which considers distinct features such as comparative analysis of protein sequences from different species, physicochemical characteristics of the exchanged amino acids and mapping of residues replacement to available 3D structures. Results are assessed as a quantitative value (a probability of being deleterious) and as a qualitative feature (benign, possibly damaging, or probably damaging). SIFT program predicts the potential pathogenic effects of amino acid substitutions on the basis of sequence homology and physical properties of the exchanged residues. Scores lower than 0.05 predict a potential deleterious effect.

Functional Analysis of GAA Mutations

An expression vector (pSHAG2), containing the wild-type *GAA* open reading frame (named GAA-Wild-Type), was used to perform site-directed mutagenesis. We introduced into this plasmid the two novel *GAA* c.1106 T > C (p.Leu369Pro) or c.2236 T > C (p.Trp746Arg) missense mutations found in Colombian patients. These constructs were named GAA-Leu369Pro and GAA-Trp746Arg, respectively. Similarly, we created four additional constructs (GAA-Trp746X, GAA-Trp746Cys, GAA-Trp746Gly and GAA-Trp746Ser) carrying mutant *GAA* versions, which represent previously reported mutations

located at position 746. The integrity of the resulting mutant constructs was, in each case, confirmed by direct sequencing.

HEK 293 T cells were seeded into 24-well plates and grown overnight in DMEM medium supplemented with 10% of fetal bovine serum, 50U/mL of penicillin and 50 μg/mL of streptomycin, in a 10% carbon dioxide and 90% air humidified incubator. Cells at 80–90% of confluence were transfected with 1.4 μg of GAA-WT or mutant constructs using polyethyleneimine. Mock transfected cells served as negative controls. Seventy-two hours after transfection, cells were washed with PBS and harvested with lysis buffer (50 mM Tris–HCl pH 7.0, 150 mM NaCl, 50 mM NaF, and 1% TritonX-100). After centrifugation (10,000 g for 10 min), the supernatant fraction was recovered. GAA activity was measured in both medium and cell homogenates (Müller-Felber et al. 2007). As described by Kroos et al. (2008), the mutation severity scoring system is based on the assessment of GAA activity levels in the medium and in the cells, and on the quality and quantity of the different molecular species that arise during GAA posttranslational modification (Kroos et al. 2008). To visualize protein biosynthesis and post-translational processing, cell homogenates and immunoprecipitated GAA from the medium were subjected to SDS-PAGE followed by Western-blotting (Müller-Felber et al. 2007). To visualize GAA on the blots we used GAA-specific polyclonal mouse and rabbit antisera as primary antibodies and goat anti-mouse IRDye 800LT (LI-COR Biosciences) and goat anti-rabbit IRDye 700LT (LI-COR Biosciences) as secondary antibodies. Transfection assays and GAA measurements were performed three times as duplicates.

Results

GAA Mutation Detection and In Silico Analysis

Sequence analysis of the complete coding region of *GAA*, performed in 11 Colombian PD patients, revealed six sequence variants previously related with PD pathogenesis: c-32-13 T > G, c.-32-2A > G, c.525delT (p.Glu176fsX45), c.1064 T > C (p.Leu355Pro), c.2481 + 102_2646del (p.Gly828_Asn882del), and c.2560 C > T (p.Arg854X) (Table 1). Pt 2 was homozygous for the p.Leu355Pro mutation and pt 3 showed compound heterozygosity for the two pathogenic variants c.-32-13 T > G and c.525delT (p.Glu176fsX45). Three related patients (pt 4, 5, and 6) shared the c.-32-2A > G/ c.2481 + 102_2646del (p.Gly828_Asn882del) compound heterozygous genotype. Pt 7 was compound heterozygote for the sequence variants c.2560 C > T (p.Arg854X) and c.-32-13 T > G. Pt 8 displayed c.-32-13 T > G homozygosity. Patients 9 and 10 were also related. They shared the c.2560 C > T (p.Arg854X) heterozygous mutation. Pt 7, 9, and 10 also carried non-synonymous and intronic variants, which were previously described in PD patients but were not related with the disease pathogenesis (Table 1).

Pt 1 and 11 displayed novel c.1106 T > C (p.Leu369-Pro) and c.2236 T > C (p.Trp746Arg) heterozygous variants, respectively. These mutations are located in the catalytic $(\beta/\alpha)_8$ barrel (p.Leu369Pro) and in the proximal C-terminal (p.Trp746Arg) domains of GAA. Both these patients are compound heterozygous for deleterious *GAA* mutations since they also presented c.1064 T > C (p.Leu355Pro) (pt 1) and c.2560 C > T (p.Arg854X) (pt 11).

At the protein level, comparative in silico analysis of the novel p.Leu369Pro and p.Trp746Arg mutations showed a strict conservation among vertebrate species of both Leucine and Tryptophan residues at positions 369 and 746, respectively (Fig. 1). Polyphen bioinformatic tool predicted that these mutations are probably damaging. Similarly, SIFT software showed probabilistic scores compatible with a potential deleterious effect (p.Leu369Pro =0.00, p.Trp746Arg =0.01).

Functional Characterization of GAA Mutations

To further investigate the effect of p.Leu369Pro and p.Trp746Arg, both these mutations were introduced in the wild-type GAA cDNA by site directed mutagenesis and transiently expressed in HEK-293 cells. Cells and media were analyzed for GAA content by polyacrylamide gel electrophoresis followed by western blotting. The results are shown in Fig. 2a (cells) and 2b (media). Mock transfected HEK-293 cells showed faint signals in cells and culture medium (Fig. 2a–b). Since HEK-293 cells constitutively express human GAA, these signals can be considered as background staining.

Cells transfected with the GAA wild-type construct contained three molecular species representing the 110 kD precursor, the 95 kD partially processed intermediate and the 76 kD mature form of GAA. The medium contained only the 110 kD GAA precursor (Fig. 2). Similar results in cells and media were obtained after transfection of HEK-293 cells with the mutated p.Trp746Cys and p.Trp746Ser constructs. Quite different results were obtained after transfection with p.Trp746Arg, p.Trp746Gly, and p.Leu369P: only the intracellular 110 kD GAA precursor and no other forms of GAA were detected in cells or media, except for possibly a little bit of 76 kD mature enzyme after transfection with p.Trp746Arg. Transfection with p.Trp746X resulted in the formation of a unique molecular species with an apparent molecular mass between 76 and 95 kD in the cells. In all cases we also measured the GAA

activity in cells and media and used the scoring system as described by Kroos et al. (2008) to evaluate the severity of all the different mutations. The results are summarized in Table 2.

Discussion

In an effort to delineate the clinical and molecular features of Colombian PD patients, we identified 11 cases in which we performed *GAA* genotype analysis. Among the six different mutations that we identified in this study and that were previously related to PD pathogenesis, two affect the *GAA* mRNA splicing (c.-32-13 T > G and c.-32-2A > G), one is a missense mutation (c.1064 T > C/p.Leu355Pro), one is a single base pair deletion (c.525delT/p. Glu176fsX45), one is a large deletion including exon 18 (c.2481 + 102_2646del/p.gly828_Asn882del), and one is a nonsense mutation (c.2560 C > T/p.Arg854X).

GAA c.-32-13 T > G was most frequently encountered since three patients (pt 3, 7, and 8) (allelic frequency = 0.27) were found to be either homozygous or heterozygous for this mutation. This sequence variation is the most common pathogenic *GAA* mutation among Caucasian individuals (>70%) affected by slowly progressive PD (Hirschhorn et al. 2001; Ko et al. 1999). Pt 3 and 7, respectively, presented p.Glu179fsX45 and p.Arg854X as second pathogenic mutation. In both patients, limb girdle weakness and orthopnea were recorded as the first PD symptoms, but the age of onset differed substantially (12 vs 27 years). These findings fit with the notion that the clinical picture of patients carrying the c.-32-13 T > G mutation can vary in terms of the age of onset and rate of disease progression due to modifying factors (Huie et al. 1994; Kroos et al. 2007). Pt 8, who displayed c.-32-13 T > G homozygosity, showed first symptoms at the age of 38 years and had mild clinical features. Homozygosity for the c.-32-13 T > G variant is particularly rare as only three cases have been reported so far (Müller-Felber et al. 2007; Laforêt et al. 2000; Labrousse et al. 2010). Similarly to the patient reported by Laforet et al., this individual was classified as a late-onset case since the first symptoms manifested after the age of 38 years. It has been proposed that the rare finding of affected c.-32-13 T > G homozygotes is probably related to the high level of residual GAA activity that is associated with this genotype. Notably, all our patients having the c.-32-13 T > G mutation shared six additional sequence variants (SNP-IDs: rs17410539, rs11150843, rs7225049, rs3176968, rs1042397, rs1042397), which together mark the most common c.-32-13 T > G haplotype encountered among Caucasian PD patients (Müller-Felber et al. 2007; Kroos et al. 2008).

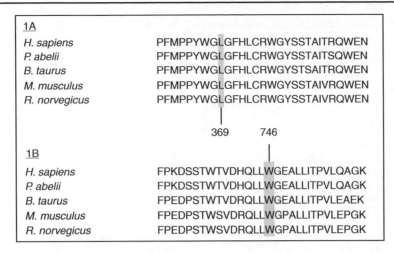

Fig. 1 Alignment of the *GAA* sequences of selected vertebrates: for leucine at position L369 (*1A*) and for tryptophane at position W746 (*1B*)

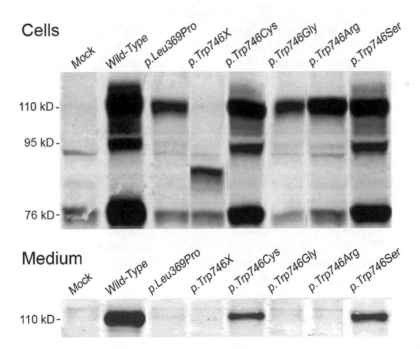

Fig. 2 Western blot analysis of GAA expression in transiently transfected HEK293 cells. Cells and culture media were harvested 72 h after transfection and the different molecular species representing precursor GAA (110 kD), partially processed precursor (95 kD) and mature GAA (76 kD) were separated by SDS-PAGE and visualized by immunoblotting as described in Materials and Methods. *Panel A*, cell homogenates; *Panel B*, media

Pt 2, who was diagnosed at the age of 11 years and homozygous for c.1064 T > C (p.Leu355Pro), confirms the previously established genotype-phenotype correlation for this mutation in that it is associated with an early childhood presentation of PD (Labrousse et al. 2010; Montalvo et al. 2004). Pt 4, 5, and 6, who are related and were diagnosed at ages of 33, 35, and 36 years, appeared to have the c.-32-2A > G/c.2481 + 102_2646del (p. Gly828_Asn882del) genotype. Interestingly, c.-32-2A > G was previously reported to be associated with early-onset

PD (Kroos et al. 2008). Thus, our results do not correlate with those previously reported and argue in favor of a relatively mild effect of the c.-32-2A > G mutation or the impact of modifying factors.

The p.Arg854X mutation, which has been reported as the most frequent GAA sequence variation among Afro-American PD patients, was identified in four of our Colombian patients (pt 7, 9, 10, and 11) at heterozygous state (Raben et al. 1999). Homozygotes for this mutation have infantile onset PD (Laforêt et al. 2000; Kroos et al.

Table 2 *GAA* sequence variations analyzed by transient expression in HEK293 cells

Nucleotide change	Amino acid change	M 110*	C110*	C95*	C76*	M%	C%	Class*
c.1106 T > C	**p.Leu369Pro**	1,1	3,4	2,4	?	0.2	3.1	B
c.2238 G > A	p.Trp746X	1,1	3,2	1,1	?	−0.2	−0.14	A
c.2238 G > C	p.Trp746Cys	3,4	3,4	3,4	3,4	5.4	29.4	D
c.2236 T > G	p.Trp746Gly	1,1	3,4	2,4	?	0.2	2.1	B
c.2236 T > C	**p.Trp746Arg**	1,1	3,4	2,4	?	−0.1	3.5	B
c.2237 G > C	p.Trp746Ser	3,4	4, 4	3,4	3,4	8.2	40.1	E

The two new missense mutations that we analyzed in this study are shown in bold

*M110, C10, C95, and C76 stand for the various molecular forms of GAA that arise during synthesis and posttranslational modification and that can be visualized by western blotting as illustrated in Fig 2a and b. The numbers refer to the severity rating system as published by Kroos et al. (2008). Class A mutations are very severe, class B mutations are potentially less severe, class D mutations are mild, and class E mutations are probably nonpathogenic. M% stands for the percentage of GAA activity in the culture medium and C% for the percentage of GAA activity in the cells as compared to Wild-Type GAA activity.

The question mark (?) signifies that there remains uncertainty about the formation of the 76 kD form of GAA

2008; Reuser et al. 1689). The patients in our study lacked cardiomyopathy and presented their first clinical signs after 10 years of age. The difference in age of onset must be due to residual GAA activity conferred by a less severe mutation located on the second allele. Indeed, pt 7 carried the c.-32-13 T > G mutation, which displays significant residual activity that apparently compensates the strong deleterious effect of the Arg854X mutation. Unfortunately, we failed to establish a more precise genotype-phenotype correlation in pt 9 and 10 (who are related) since we did not find the second *GAA* mutation.

Pt 11 with infantile PD had the first clinical signs (hypotonia, delayed motor development) at 3 months of age. Apart from the p.Arg854X amino acid change, this patient displayed the novel p.Trp746Arg mutation, which is located in the proximal C-terminal domain of GAA. Comparative sequence analysis of vertebrate species demonstrated a strict conservation of this tryptophane suggesting its essential functional role. The substitution Trp to Arg implicates a drastic modification in terms of physicochemical properties. Indeed, tryptophane is a nonpolar aromatic amino acid whereas Arg is a small polar hydrophilic residue. These features were reinforced by SIFT and Polyphen2 bioinformatic tools which predicted a potential deleterious effect of the p.Trp746Arg mutation.

In accordance with these predictions, transient expression studies demonstrated that the structural changes conferred by p.Trp746Arg do not inhibit the synthesis of the 110 kD precursor but affect the processing and intracellular transport of GAA (Fig. 2). According to the mutation severity scoring system proposed by Kroos et al. (2008) these results argue in favor of a potentially less severe mutation. Additional analysis of previously described variants revealed similar results for the substitu-

tion p.Trp746Gly, which was found during a newborn screening program in a patient with low GAA activity (Labrousse et al. 2010). The p.Trp746Cys and p.Trp746Ser variants can both be classified as relatively mild mutations since the 110 kD GAA precursor as well as processed forms of GAA were detected in both cases, albeit in less than normal amount (Fig. 2a–b, Table 2). Notably, up till now some controversy existed concerning the pathogenicity of p.Trp746Cys since it was described as a polymorphism as well as a causal mutation (Wan et al. 2008; Chien et al. 2011). Our results argue in favor of a mildly deleterious effect.

Transient expression of the p.Trp746X mutation resulted in the appearance of a truncated precursor, which is apparently stable enough to be visualized by western blotting, but lacks catalytic activity. This situation corroborates previous clinical findings in which patients carrying this mutation are affected by infantile PD (Kishnani et al. 2006; Beesley et al. 1998).

Next to p.Trp764Arg, p.Leu369Pro was the second novel mutation identified in our study. It was found in an affected child in combination with p.Leu355Pro. Pulmonary distress was diagnosed at 1 year of age, and the patient required ventilation support at the age of 9 years. The muscle weakness was especially severe. In silico analysis of this mutation suggested a pathogenic effect, similar to p.Trp746Arg, based on the strict conservation of the Leu residue at position 369 among vertebrate species. This prediction was validated by transient expression studies since the amino acid substitution appeared to hamper the posttranslational modification and intracellular transport of GAA (Fig. 2a–b, Table 2).

Finally, from an ethnical point of view, it is interesting that the two most common *GAA* mutations found in our study are also common in Caucasian (c.-32-13 T > G) and

African (p.Arg854X; allele frequency = 0.43) populations (Becker et al. 1998). Notably, all Colombian patients presenting p.Arg854X shared a previously identified haplotype found in black PD patients from the United States, the Ivory Coast, Ghana, and Namibia (Becker et al. 1998; Hermans et al. 1993). These findings evoke a remarkable Afro-American and European ancestry of the Colombian population.

In summary, we investigated the genetics of PD in the Colombian population and identified two novel causative mutations in the GAA gene in addition to other previously reported pathogenic sequence variations. Valuable information on the genotype-phenotype correlation was obtained that is expected to facilitate and improve genetic counseling of affected individuals and their families.

Acknowledgments This work was supported by the Universidad del Rosario (Grant CS/Genetics) and by Genzyme Corporation, Colombia.

References

Becker JA et al (Apr 1998) The African origin of the common mutation in African American patients with glycogen-storage disease type II. Am J Hum Genet 62(4): 991–994. ISSN 0002-9297. http://www.ncbi.nlm.nih.gov/pubmed/9529346

Beesley CE, Child AH, Yacoub MH (1998) The identification of five novel mutations in the lysosomal acid a-(1-4) glucosidase gene from patients with glycogen storage disease type II. Mutations in brief no. 134. Hum Mutat 11(5): 413. ISSN 1059-7794. http://www.ncbi.nlm.nih.gov/pubmed/10206684

Boerkoel CF et al (Apr 1995) Leaky splicing mutation in the acid maltase gene is associated with delayed onset of glycogenosis type II. Am J Hum Genet 56(4): 887–897. ISSN 0002-9297. http://www.ncbi.nlm.nih.gov/pubmed/7717400

Chien YH et al (Jun 2011) Later-onset Pompe disease: early detection and early treatment initiation enabled by newborn screening. J Pediatr 158(6): 1023–1027.e1. ISSN 1097-6833. http://www.ncbi.nlm.nih.gov/pubmed/21232767

Den Dunnen JT, Antonarakis SE (2000) Mutation nomenclature extensions and suggestions to describe complex mutations: a discussion. Hum Mutat 15(1): 7–12. ISSN 1059-7794. http://www.ncbi.nlm.nih.gov/pubmed/10612815

Hermans MM et al (1993) Two mutations affecting the transport and maturation of lysosomal alpha-glucosidase in an adult case of glycogen storage disease type II. Hum Mutat 2(4): 268–273. ISSN 1059-7794. http://www.ncbi.nlm.nih.gov/pubmed/8401535

Hirschhorn R, Reuser A (2001) Glycogen storage disease type II (GSDII). In: Scriver C, Beaudet A, Sly W, Valle D (eds) The metabolic and molecular bases of inherited disease. McGraw-Hill, New York, pp 3389–3420

Huie ML et al (Dec 1994) Aberrant splicing in adult onset glycogen storage disease type II (GSDII): molecular identification of an IVS1 (-13 T- > G) mutation in a majority of patients and a novel IVS10 (+1GT- > CT) mutation. Hum Mol Genet 3(12): 2231–2236. ISSN 0964-6906. http://www.ncbi.nlm.nih.gov/pubmed/7881425

Kishnani PS et al (May 2006) A retrospective, multinational, multicenter study on the natural history of infantile-onset Pompe disease. J Pediatr 148(5): 671–676. ISSN 0022-3476. http://www.ncbi.nlm.nih.gov/pubmed/16737883

Ko TM et al (1999) Molecular genetic study of Pompe disease in Chinese patients in Taiwan. Hum Mutat 13(5): 380–384. ISSN 1059-7794. http://www.ncbi.nlm.nih.gov/pubmed/10338092

Kroos MA et al (Jan 2007) Broad spectrum of Pompe disease in patients with the same c.-32-13 T- > G haplotype. Neurology 68(2): 110–115. ISSN 1526-632X. http://www.ncbi.nlm.nih.gov/pubmed/17210890

Kroos M et al (Jun 2008) Update of the Pompe disease mutation database with 107 sequence variants and a format for severity rating. Hum Mutat 29(6): E13–26. ISSN 1098-1004. http://www.ncbi.nlm.nih.gov/pubmed/18425781

Labrousse P et al (Apr 2010) Genetic heterozygosity and pseudodeficiency in the Pompe disease newborn screening pilot program. Mol Genet Metab 99(4): 379–383. ISSN 1096-7206. http://www.ncbi.nlm.nih.gov/pubmed/20080426

Laforêt P et al (Oct 2000) Juvenile and adult-onset acid maltase deficiency in France: genotype-phenotype correlation. Neurology 55(8): 1122–1128. ISSN 0028-3878. http://www.ncbi.nlm.nih.gov/pubmed/11071489

Li Y et al (Oct 2004) Direct multiplex assay of lysosomal enzymes in dried blood spots for newborn screening. Clin Chem 50(10): 1785–1796. ISSN 0009-9147. http://www.ncbi.nlm.nih.gov/pubmed/15292070

Montalvo AL et al (Mar 2004) Glycogenosis type II: identification and expression of three novel mutations in the acid alpha-glucosidase gene causing the infantile form of the disease. Mol Genet Metab 81(3): 203–208. ISSN 1096-7192. http://www.ncbi.nlm.nih.gov/pubmed/14972326

Müller-Felber W et al (Oct 2007) Late onset Pompe disease: clinical and neurophysiological spectrum of 38 patients including long-term follow-up in 18 patients. Neuromuscul Disord 17(9–10): 698–706. ISSN 0960-8966. http://www.ncbi.nlm.nih.gov/pubmed/17643989

Oba-Shinjo SM et al (Nov 2009) Pompe disease in a Brazilian series: clinical and molecular analyses with identification of nine new mutations. J Neurol 256(11): 1881–1890. ISSN 1432-1459. http://www.ncbi.nlm.nih.gov/pubmed/19588081

Palmer RE et al (Jan 2007) Pompe disease (glycogen storage disease type II) in Argentineans: clinical manifestations and identification of 9 novel mutations. Neuromuscul Disord 17(1): 16–22. ISSN 0960-8966. http://www.ncbi.nlm.nih.gov/pubmed/17056254

Pittis MG et al (Jun 2008) Molecular and functional characterization of eight novel GAA mutations in Italian infants with Pompe disease. Hum Mutat 29(6): E27–36. ISSN 1098-1004. http://www.ncbi.nlm.nih.gov/pubmed/18429042

Raben N et al (1999) Novel mutations in African American patients with glycogen storage disease Type II. Mutations in brief no. 209. Hum Mutat 13(1): 83–84. ISSN 1059-7794. http://www.ncbi.nlm.nih.gov/pubmed/10189220

Raben N et al (Nov–Dec 2007) Deconstructing Pompe disease by analyzing single muscle fibers: to see a world in a grain of sand... Autophagy 3(6):546–552. ISSN 1554-8627. http://www.ncbi.nlm.nih.gov/pubmed/17592248.

Reuser AJ et al (Jun 1987) Clinical diversity in glycogenosis type II. Biosynthesis and in situ localization of acid alpha-glucosidase in mutant fibroblasts. J Clin Invest 79(6): 1689–1699. ISSN 0021-9738. http://www.ncbi.nlm.nih.gov/pubmed/3108320

Sharma MC et al (2005) Delayed or late-onset type II glycogenosis with globular inclusions. Acta Neuropathol 110(2): 151–157. ISSN 0001-6322. http://www.ncbi.nlm.nih.gov/pubmed/15986226

Slonim AE et al (Jan 2007) Modification of the natural history of adult-onset acid maltase deficiency by nutrition and exercise therapy. Muscle Nerve 35(1): 70–77. ISSN 0148-639X. http://www.ncbi.nlm.nih.gov/pubmed/17022069

Sugawara K et al (Jun 2009) Structural modeling of mutant alpha-glucosidases resulting in a processing/transport defect in Pompe disease. J Hum Genet 54(6): 324–330. ISSN 1435-232X. http://www.ncbi.nlm.nih.gov/pubmed/19343043

van den Hout HM et al (Aug 2003) The natural course of infantile Pompe's disease: 20 original cases compared with 133 cases from the literature. Pediatrics 112(2): 332–340. ISSN 1098-4275. http://www.ncbi.nlm.nih.gov/pubmed/12897283

van der Ploeg AT, Reuser AJ (Oct 2008) Pompe's disease. Lancet 372(9646): 1342–1353. ISSN 1474-547X. http://www.ncbi.nlm.nih.gov/pubmed/18929906

van der Ploeg A et al (2010) A randomized study of alglucosidase alfa in late-onset Pompe's disease. N Engl J Med 362(15):1396–1406. http://www.ncbi.nlm.nih.gov/pubmed/20393176

Wan L et al (Jun 2008) Identification of eight novel mutations of the acid alpha-glucosidase gene causing the infantile or juvenile form of glycogen storage disease type II. J Neurol 255(6): 831–838. ISSN 0340-5354. http://www.ncbi.nlm.nih.gov/pubmed/18458862

JIMD Reports
DOI 10.1007/8904_2012_140

CASE REPORT

Successful Live Birth following Preimplantation Genetic Diagnosis for Phenylketonuria in Day 3 Embryos by Specific Mutation Analysis and Elective Single Embryo Transfer

Stuart Lavery · Dima Abdo · Mara Kotrotsou ·
Geoff Trew · Michalis Konstantinidis · Dagan Wells

Received: 11 December 2011 /Revised: 15 February 2012 /Accepted: 5 March 2012 /Published online: 31 March 2012
© SSIEM and Springer-Verlag Berlin Heidelberg 2012

Abstract Phenylketonuria (PKU) is an autosomal recessive inherited metabolic disorder caused by a complete or near-complete deficiency of the liver enzyme phenylalanine hydroxylase (PAH), which converts the amino acid phenylalanine to tyrosine, leading to the increase of blood and tissue concentration of phenylalanine to toxic levels. PKU is not life threatening but is treated through lifelong dietary management. If untreated, it can lead to severe learning disability, brain function abnormalities, behavioural and neurological problems. The non-life threatening nature of PKU has until now caused some debate on whether to licence its detection by preimplantation genetic diagnosis (PGD). We report the first successful live birth in the UK following single cell embryo biopsy and PGD for the detection of two different mutations in the (PAH) gene. This case highlights both an important scientific development as well as the ethical challenge in offering couples who carry PKU this new reproductive option when starting their family.

Communicated by: John H Walter

Competing interests: None declared

S. Lavery (✉) · D. Abdo · M. Kotrotsou · G. Trew
IVF Hammersmith, Hammersmith Hospital, Du Cane Road,
London W12 0HS, United Kingdom
e-mail: stuart.lavery@imperial.ac.uk

M. Konstantinidis · D. Wells
Reprogenetics UK, Institute of Reproductive Sciences, Oxford
Business Park, North Oxford, Oxfordshire
OX4 2HW, United Kingdom

M. Konstantinidis · D. Wells
Nuffield Department of Obstetrics & Gynaecology,
John Radcliffe Hospital, University of Oxford, Oxford OX3 9DU,
United Kingdom

Introduction

Phenylketonuria (PKU) is an inherited metabolic disorder that is transmitted as an autosomal recessive trait. Individuals with homozygous mutations in the phenylalanine hydroxylase (PAH) gene lack the enzyme PAH that is essential for the breakdown of the amino acid phenylalanine. Accumulation of phenylalanine in the body can cause damage to the central nervous system and subsequently cognitive and behavioural abnormalities and mental retardation in both children and adults (Mitchell et al. 2011; Enns et al. 2010; Feillet et al. 2010a).

The incidence of PKU varies in different populations, with a prevalence of 1 in 10,000 births in Europe and the United States (Dobrowolski et al. 2011; Blau et al. 2010; Hardelid et al. 2007). Its prevalence is higher in northern Europeans and it has an incidence of 1 in 4,500 in the Irish (Blau et al. 2010; Magee et al. 2002).

Treatment consists of a diet low in phenylalanine starting shortly after birth and continued into adulthood ('diet for life') (Feillet et al. 2010a). If the disorder is untreated it can result in severe mental retardation by the end of the first year of life. Screening of newborns immediately after birth is standard practice in most developed countries (van Spronsen 2010).

Couples who both carry mutations are faced with difficult decisions when considering starting or enlarging their family. Some couples will accept a one in four chance of an affected pregnancy. Others will choose to conceive naturally and then undergo more conventional invasive prenatal diagnoses such as chorionic villus sampling at around 11 weeks or amniocentesis at 16 weeks. These procedures carry around a 1% risk of pregnancy loss. If a pregnancy is diagnosed as affected by PKU couples will then face the option of either terminating the pregnancy or continuing the pregnancy knowing the child will need lifelong management. Preimplantation genetic diagnosis represents a further reproductive option for these couples. Embryos are produced by the in vitro fertilisation (IVF) process and can be analysed for the presence of the affected genes. Embryos free of the disease can then be transplanted back into the uterus. This allows the couple the reassurance from the very beginning of pregnancy that the child will be unaffected by PKU and enables them to avoid invasive prenatal diagnosis. PGD was first performed in our unit over 20 years ago (Handyside et al. 1989). During this time PGD for lethal disorders has become more widely accepted, but there remains some debate about the application of this technology to screen for more chronic disorders where effective management is available.

In the United Kingdom licences to perform PGD for any disease are granted by the UK Human Fertilisation and Embryology Authority (HFEA). Our case was the first case of PGD for PKU performed in the United Kingdom and a licence was granted by the HFEA following formal application.

Materials and Methods

A couple presented to our clinic for PGD after their first child was diagnosed with PKU. The mother was 31 years old and a carrier of a mutation in the PAH gene (c.1241 A > G). The father was a carrier of a different mutation in the PAH gene (c.194 T > C). This presented a one in four risk of an affected pregnancy. The couple were offered the option of conventional prenatal screening but declined and were keen on the option of PGD. A PGD workup was performed initially on the parents' DNA and then tested on single cells. For direct detection of the mutations carried by the couple, primers were designed using the Primer3 software (http://frodo.wi.mit.edu/primer3) with reference to the human phenylalanine hydroxylase gene sequence (Genbank: NG_008690.1). The following primers for DNA amplification and minisequencing were designed:

c.194 T > C_F (5'-ACCCTCCCCATTCTCTCTTC-3'),
c.194 T > C_R (5'-AGGCAGGCTACGTTTATCCA-3'),
c.194 T > C_Minisequencing_F (5'-TGATG-TAAACCTGACCCACA-3'),
c.1241A > G_F (5'-GTGGTTTTGGTCTTAGGAA CTTTG-3'),
c.1241A > G_R (5'-ATCTTAAGCTGCTGGGTAT TGT-3'),
c.1241A > G_Minisequencing_F (5'-TCGGCCCTTCT CAGTTCGCT-3').

The intragenic STR-3 and VNTR-13 polymorphic markers were also used in the PGD protocol (Verlinsky et al. 1999).

The STR-3 forward primer was 5'-fluorescently labelled with 6-carboxyfluorescein (6- FAM), while the VNTR-13 forward primer was labelled with 6-carboxyhexafluorescein (HEX).

The female patient then underwent a cycle of IVF. A long day 21 suppression protocol was used with a gonadotropin releasing hormone (GnRH) agonist. After 10 days of stimulation with 150 IU of recombinant follicle stimulating hormone (FSH), oocyte release was triggered with 0.25 mg of recombinant human chorionic gonadotropin (hCG). This was followed by ultrasound-guided trans-vaginal oocyte retrieval, performed under sedation. Sixteen oocytes were retrieved, of which fifteen were suitable for intracytoplasmic sperm injection (ICSI). Thirteen oocytes were fertilised normally and were cultured to day 3. On day 3 of development, 12 embryos had at least five blastomeres and were suitable for biopsy. A hole was performed in the zona pellucida using laser (Saturn™, Research Instruments) and a single blastomere was removed from each embryo using Humagen biopsy micropipettes. Each blastomere was washed in NWB (non-stick washing buffer) (Reprogenetics UK, Oxford, UK) then placed in a sterile 0.2 mL PCR tube in 1 μL of clean NWB and placed on ice to be sent to Reprogenetics UK Ltd for genetic analysis. Following the biopsy the embryos were cultured individually to day 5 in numerically labelled 4-well dishes to prevent cross-contamination and to allow the identification of each embryo.

Biopsied cells were lysed and amplified using the SurePlex DNA Amplification System (Rubicon, USA) according to manufacturer's instructions. Subsequently, each of the four loci (two mutation sites and two polymorphic markers) was amplified in singleplex reactions. For the amplification of the sequences encompassing the two mutations the HotMaster *Taq* DNA polymerase kit (5 Prime, UK) was used. Reaction mixtures contained PCR grade water, 1x HotMaster *Taq* Buffer (with 25 mM Mg^{2+}), dNTPs (200 μM each), 0.8 μM each primer, 0.6 units HotMaster *Taq* DNA polymerase and 1 μl of SurePlex amplified product for a final volume of 15 μL. Thermal cycling consisted of an initial denaturation step of 96°C for 1 min, followed by 45 cycles of 94°C for 15 s, 54.5°C for 15 s, and 65°C for 45 s, then a final extension step of 65°C for 2 min. For amplification of the microsatellite markers the Expand Long Template PCR System was used (Roche Diagnostics Ltd., UK). Reaction mix contained PCR grade water, 1x Expand Long Template Buffer 3 (with 27.5 mM MgCl$_2$ and detergents), dNTPs (350 μM each), 2 μM each primer, 1.5 units Expand Long Template enzyme mix and 1 μL of SurePlex amplified product for a final volume of 15 μL. Thermal cycling was carried out for 45 cycles as described by Kakourou et al. 2010 (annealing temperature was 54.5°C). After amplification of the sequences involving the two mutations, the amplified products were cleaned using EXOSAP-IT (Affymetrix, UK) and minisequencing was carried out using the SNaPshot Multiplex kit (Applied Biosystems, UK) according to manufacturer's instructions. Analysis of the minisequencing products and amplified microsatellite products was accomplished through capillary electrophoresis by using a 3130 genetic analyser (Applied Biosystems). Analysis of the data was carried out using GeneMapper v4.0 software (Applied Biosystems).

Results

Embryo status was based upon interrogation of each of the mutation sites in the PAH gene and also analysis of the inheritance of alleles from two intragenic polymorphisms. The combination of direct mutation detection and linkage analysis provides multiple opportunities to detect affected embryos, a diagnostic redundancy that is critical for accurate analysis of single cells. Following genetic analysis, three embryos were found to be free of both parental mutations and therefore unaffected, four were found to be carriers of the maternal mutation, four were affected with PKU (both mutations detected) and one embryo failed to yield any result (Fig. 1). All embryos were cultured to the blastocyst stage and one hatching unaffected blastocyst of grade 5Cc (Gardner blastocyst grading system) (Gardner et al. 2000) was transferred on day 5 of development, using a soft embryo transfer catheter (Rocket® Embryon® SOFT ET catheter) under ultrasound guidance. This resulted in a clinical pregnancy and subsequently the live birth of a healthy baby boy, as confirmed by post-natal testing. The remaining two unaffected blastocysts and three of the four carriers were cryopreserved.

Discussion

In the UK, fertility clinics require the approval of the HFEA prior to performing PGD for a genetic condition. The HFEA must therefore agree that a particular genetic condition is sufficiently serious to warrant using PGD. Once a condition is approved it is added to the list of HFEA licenced PGD conditions and approval on an individual basis is no longer required.

The HFEA Code of Practice – Guidance on embryo testing states that: *Preimplantation genetic diagnosis (PGD) can be carried out for a heritable condition only in two circumstances: a) where there is a particular risk that the embryo to be tested may have a genetic, mitochondrial or chromosomal abnormality, and the Authority is satisfied that a person with the abnormality will have or develop a serious disability, illness or medical condition, or b) where there is a particular risk that any resulting child will have or develop a gender related serious disability, illness or medical condition.* It further states that the centre performing the PGD should consider several factors when deciding if PGD is appropriate or not, one of which is *the availability of effective therapy, now and in the future* (HFEA Code of Practice 2011).

PKU is a chronic genetic condition that requires life-long management in order to prevent the occurrence of learning and behavioural disability and mental retardation. The condition is not life threatening, albeit it requires chronic treatment. The life-long requirement of treatment and the implications this may have on the affected individual's life, were the determining factors in approving PKU for PGD in the UK.

The licensing of PGD for PKU in the UK represents a major ethical shift in the application of PGD for chronic treatable conditions. This was evident during our process of applying for HFEA approval for testing for this condition. The first application was rejected on the grounds that PKU is a treatable non-life-threatening condition and that differences in outcomes exist between those whose condition is detected at birth and those whose condition is detected at a later stage causing more serious consequences. The HFEA also required more evidence of the effect of the condition on the quality of life of affected individuals and their families. Our application was reviewed and approved following correspondence from the National Society of Phenylketonuria in the UK (www.nspku.org), a patient support organisation which is a member of Genetic Alliance UK (www.geneticalliance.org.uk), a national charity supporting those affected by genetic disorders and their families. This correspondence outlined the adverse effects of PKU on the quality of life of patients and their caregivers, especially in childhood when a strict diet is required until the age of 10 years, and in adolescence when the diet compliance proves to be more difficult. In adulthood, self-management of the condition becomes quite challenging and poses great difficulties both psychologically and socially (Feillet et al. 2010a; Macdonald et al. 2010). The cost, inconvenience, and sometimes difficult access to PKU diet increase the difficulty of adhering to it. Although there is still debate over whether adult PKU patients need a strict phenylalanine-restricted diet and despite the fact that there are some adults with PKU who do not follow a restricted diet and lead normal lives, many studies have advocated following a restricted diet in adulthood as it has been shown that high phenylalanine levels in adults can have adverse effects on mood, sustained attention, concentration, and cognitive ability (Feillet et al. 2010b; Ten Hoedt et al. 2011; Macdonald et al. 2010; Moyle et al. 2007; Simon et al. 2008). Restricted diet is also crucial in female adults with PKU planning for a

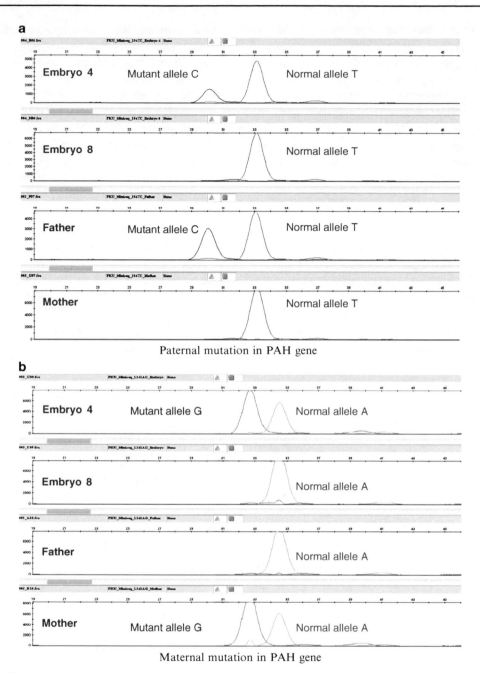

Fig. 1 Results of PGD showing (**a**) the paternal mutation and (**b**) the maternal mutation. Embryo 4 was affected (both mutations present) and embryo 8 was unaffected (no mutations present)

pregnancy, as maternal PKU has been shown to cause fetal abnormalities including microcephaly, congenital heart disease, and mental retardation (Prick et al. 2012). Moreover, it has been shown in a systemic literature review by Enns et al. (2010) that even with early detection and management of PKU through diet, the neurocognitive, psychosocial, quality of life, growth, nutrition, and bone pathology outcomes in these patients were suboptimal. Individuals with PKU require lifelong monthly blood tests to ensure that blood phenylalanine levels are not elevated above the recommended values ranging from 120 to 360 μM/L

up to 12 years of age and from 120 to 900 μM/L after the age of 12, with national guidelines varying from one country to another (Ahring et al. 2011; Anastasoaie et al. 2008; Enns et al. 2010; van Spronsen 2010). Moreover, fluctuation in blood phenylalanine levels and not just the absolute values is equally important and can also affect cognitive functioning in PKU patients (Anastasoaie et al. 2008; Feillet et al. 2010b).

In addition to a diet low in phenylalanine which is present in all protein-rich foods, dietary supplementation, such as protein-free formulas and protein substitutes, is also required in combination with the diet (Feillet et al. 2010b; Macdonald et al. 2010). Other non-traditional therapies are also being used such as large neutral amino acids (LNAA) which have shown to be beneficial in maintaining low phenylalanine levels in the body without the need of dietary control (van Spronsen et al. 2010) and other therapies that are still under experimentation such as gene therapy, enzyme replacement therapy and cell transplantation (Mitchell et al. 2011; van Spronsen and Enns 2010).

The licence committee were satisfied that our revised application provided clarification on the likelihood of PKU being detected in the newborn: there is a neonatal screening programme in the UK to test all newborns at 6–14 days, but it has been shown that there are regional variations in the timing of this test and the time of starting treatment, with about 8% of newborns with severe PKU still not treated by 20 days after birth (Smith et al. 1991). After studying our revised application and correspondence from the NSPKU, the licensing committee concluded that even if the condition is detected at birth and is treated immediately, the lifelong dietary regime and medication in the child can be seriously intrusive and socially and psychologically invasive and damaging. It can be physically unpleasant, emotionally difficult, and disruptive of family life and can seriously adversely affect the quality of family and social relationships. Moreover, there is still the possibility that the affected individual will develop neurological problems or have seizures or mental retardation, and even when detected early, there is a risk of the condition not being treated sufficiently early so that in these cases more serious effects can manifest. In making this decision the HFEA licence committee authorised our centre to perform embryo testing for PKU and agreed that PKU should be added to the publicly available list of PGD conditions, allowing all licenced PGD centres in the UK to test for this condition.

Previously, PGD for PKU was performed following polar body biopsy as the mutation which was being detected was maternal (Verlinsky et al. 2001). As only the polar body from maternal oocytes was analysed no paternal input could be assessed. This case represents the first use of PGD in the UK for the diagnosis of PKU and, more specifically, the first reported case of PGD for PKU following embryo biopsy for the detection of two different mutations in the PAH gene, one maternal and the other paternal.

Assisted conception techniques often involve the transfer of multiple embryos into the uterus in an attempt to maximise success. This has led to an epidemic of multiple pregnancies. Our couple were already caring for a child affected by PKU and so a mutual decision was made to electively transfer one embryo to try and avoid the possibility of caring for both an affected child and the burden of a multiple pregnancy.

Whilst this case adds to the scientific literature for PGD and gives additional reproductive choices for couples who carry the PKU mutations, it is the ethical considerations that may have more impact. This case highlights a tension between personal reproductive choices that couples have and an interest from the state as represented by the regulatory body. Who is best placed to make these decisions? Having reversed their initial refusal to grant a licence, it is likely this could herald a change in how the HFEA regulator might view applications for other PGD cases for other non-life threatening conditions.

Acknowledgements The authors would like to thank Anastasia Mania (IVF Hammersmith) for her invaluable role during the licence application to the HFEA.

DW is supported by the NIHR Biomedical Research Centre Oxford.

Synopsis

We report the first successful live birth in the UK following single cell embryo biopsy and PGD for the detection of two different mutations in the PAH gene.

Conflict of Interest

Dagan Wells is a recipient of research grants from the Oxford NIHR Biomedical Research Centre, Gema Diagnostics and from Merck Serono. He has also received honoraria for lectures at educational meetings.

References

Ahring A, Belanger QA, Dokoupil K, Gokmen-Ozel H, Lammardo AM, MacDonald A, Motzfeldt K, Nowacka N, Robert M, van Rijn M (2011) Blood phenylalanine control in phenylketonuria: a survey of 10 European centres. Eur J Clin Nutr 65:275–278

Anastasoaie V, Kurzius L, Forbes P, Waisbren S (2008) Stability of blood phenylalanine levels and IQ in children with phenylketonuria. Mol Genet Metab 95:17–20

Blau N, van Spronsen FJ, Levy HL (2010) Phenylketonuria. Lancet 376:1417–1427

Dobrowolski SF, Heintz C, Miller T, Ellingson C, Ellingson C, Ozer I, Gokcay G, Baykal T, Thony B, Demirkol M, Blau N (2011) Molecular genetics and impact of residual in vitro phenylalanine hydroxylase activity on tetrahydrobiopterin responsiveness in Turkish PKU populations. Mol Genet Metab 102:116–121

Enns GM, Koch R, Brumm V, Blakely E, Suter R, Jurecki E (2010) Suboptimal outcomes in patients with PKU treated early with diet alone: revisiting the evidence. Mol Genet Metab 101:99–109

Feillet F, MacDonald A, Hartung D, Burton B (2010a) Outcomes beyond phenylalanine: an international perspective. Molec Genet Metab 99:S79–S85

Feillet F, van Spronsen FJ, MacDonald A, Trefz FK, Demirkol M, Giovannini M, Belanger-Quintana A, Blau N (2010b) Challenges and pitfalls in the management of phenylketonuria. Pediatrics 126:333–341

Gardner D, Lane M, Stevens J et al (2000) Blastocyst score affects implantation and pregnancy outcome: towards a single blastocyst transfer. Fertil Steril 73:1155–1158

Handyside AH, Pattinson JK, Penketh RJ, Delhanty JD, Winston RM, Tuddenham EG (1989) Biopsy of human preimplantation embryos and sexing by DNA amplification. Lancet 18:347–349

Hardelid P, Cortina-Borja M, Munro A, Jones H, Cleary M, Champion MP, Foo Y, Scriver CR, Dezateux C (2007) The birth prevalence of PKU in Populations of European, South Asian and Sub-Saharan African ancestry living in south east England. Annals Hum Genet 71:1–7

Human Fertilisation and Embryology Authority (HFEA) (2011) Code of Practice 8th Edition

Kakourou G, Dhanjal S, Mamas T, Serhal P, Delhanty JD, Sengupta SB (2010) Modification of the triplet repeat primed polymerase chain reaction method for detection of the CTG repeat expansion in myotonic dystrophy type 1: application in preimplantation genetic diagnosis. Fertil Steril 94:1674–1679

MacDonald A, Gokmen-Ozel H, van Rijn M, Burgard P (2010) The reality of dietary compliance in the management of phenylketonuria. J Inherit Metab Dis 33:665–670

Magee AC, Ryan K, Moore A, Trimble ER (2002) Follow up of fetal outcome in cases of maternal phenylketonuria in Northern Ireland. Arch Dis Child Fetal Neonatal Ed 87:F141–F143

Mitchell JJ, Trakadis YJ, Scriver CR (2011) Phenylalanine hydroxylase deficiency. Genet Med 13:697–707

Moyle JJ, Am F, Arthur M, Bynevelt M, Burnett JR (2007) Meta-analysis of neuropsychological symptoms of adolescents and adults with PKU. Neuropsychol Rev 17:91–101

Prick BW, Hop WC, Duvekot JJ (2012) Maternal phenylketonuria and hyperphenylalaninemia in pregnancy: pregnancy complications and neonatal sequelae in untreated and treated pregnancies. Am J Clin Nutr 95:374–382

Simon E, Schwarz M, Roos J, Dragano N, Geraedts M, Siegrist J, Kamp G, Wendel U (2008) Evaluation of quality of life and description of the sociodemographic state in adolescent and young adult patients with phenylketonuria (PKU). Health Qual Life Outcomes 6:25–31

Smith I, Cook B, Beasley M (1991) Review of neonatal screening programme for phenylketonuria. BMJ 303:333–335

Ten Hoedt AE, de Sonneville LMJ, Francois B, ter Horst NM, Janssen MCH, Rubio-Gozalbo ME, Wijburg FA, Hollak CEM, Bosch AM (2011) High phenylalanine levels directly affect mood and sustained attention in adults with phenylketonuria: a randomised, double-blind, placebo-controlled, crossover trial. J Inherit Metab Dis 34:165–171

Van Spronsen FJ (2010) Phenylketonuria: a 21st century perspective. Nat Rev Endocrinol 6:509–514

Van Spronsen FJ, Enns GM (2010) Future treatment strategies in phenylketonuria. Mol Genet Metab 99:S90–S95

Van Spronsen FJ, de Groot MJ, Hoeksma M, Reijngoud D, van Rijn M (2010) Large neutral amino acids in the treatment of PKU: from theory to practice. J Inherit Metab Dis 33:671–676

Verlinsky Y, Rechitsky S, Verlinsky O, Ivachnenko V, Lifchez A, Kaplan B, Moise J, Valle J, Borkowski A, Nefedova J, Goltsman E, Strom C, Kuliev A (1999) Prepregnancy testing for single-gene disorders by polar body analysis. Genet Test 3:185–190

Verlinsky Y, Rechitsky S, Verlinsky O, Strom C, Kuliev A (2001) Preimplantation testing for phenylketonuria. Fertil Steril 76:346–349

JIMD Reports
DOI 10.1007/8904_2012_141

RESEARCH REPORT

The Transforming Growth Factor-Beta Signaling Pathway Involvement in Cardiovascular Lesions in Mucopolysaccharidosis-I

S. Yano · C. Li · Z. Pavlova

Received: 27 February 2012 / Revised: 27 February 2012 / Accepted: 6 March 2012 / Published online: 18 April 2012
© SSIEM and Springer-Verlag Berlin Heidelberg 2012

Abstract Mucopolysaccharidoses (MPS) are a group of genetic disorders due to deficiency of lysosomal enzymes resulting in impaired glycosaminoglycan metabolism. All types of MPS can present with cardiovascular manifestation, although MPS-I, II, and VI seem to have more severe involvement than the other types. Enzyme replacement therapy (ERT) is available for MPS-I, II, and VI. Cardiovascular changes including hypertrophic cardiomyopathy, thickened valvular lesions, and coronary artery lesions often poorly respond to ERT and are well known as leading causes of death in patients with MPS-I. The mechanisms to cause these changes in MPS-I have not been well characterized. Immunohistopathological studies were conducted on the cardiac specimens from a patient with MPS-I who died due to sudden cardiac failure. Phosphorylated Smad2 staining showed hyperactive transforming growth factor-beta (TGF-β) signals in the intimal layer with myointimal proliferation causing stenosis in the coronary arteries as well as in the thickened endocardium and in the myocardial cells. TGF-β is involved in the pathogenesis of cardiovascular diseases including hypertrophic cardiomyopathy and vascular atherosclerosis. The primary mechanisms to cause hyperactive TGF-β signals in MPS-I are unknown. The similar mechanisms leading to hyperactive TGF-β signals may exist in the other types of MPS. The findings of TGF-β hyperactivity in the cardiovascular lesions in a patient with MPS-I may lead to a new therapeutic approach. Further studies are warranted to evaluate the effectiveness of the medications that suppress TGF-β signals, such as losartan, in preventing or improving cardiaovascular lesions in patients with MPS.

Abbreviations
MPS   Mucopolysaccharidosis
TGF-β   Transforming growth factor-beta

Communicated by: Maurizio Scarpa

Competing interests: None declared

S. Yano (✉)
Genetics Division, Department of Pediatrics, LAC + USC
Medical Center, University of Southern California, General
Laboratory Building Rm 1 G-24, 1801 Marengo Street,
Los Angeles 90033 CA, USA
e-mail: syano@usc.edu

C. Li
Research Division, Neonatology Division, Department of
Pediatrics, LAC + USC Medical Center, University of Southern
California, Los Angeles, CA, USA

Z. Pavlova
Department of Pathology, Childrens Hospital Los Angeles,
University of Southern California, Los Angeles, CA, USA

Introduction

Mucopolysaccharidoses (MPS) are a group of genetic disorders due to deficiency of lysosomal enzymes resulting in impaired glycosaminoglycan metabolism. The disorders have heterogeneous clinical phenotypes including natural history and symptoms even among patients with the same type of MPS. Systemic gradual accumulation of glycosaminoglycan in the lysosomes typically causes chronic progressive nature and often involves many organ systems, resulting in early death. Cardiovascular lesions in MPS-I have been studied more intensively compared to other types of MPS. Cardiovascular findings in autopsies on patients with MPS-I (including Hurler, Hurler-Scheie, and Scheie variants) revealed approximately 70% of valvular involvement and approximately 50% of arterial involvement including coronary artery stenosis (Krovetz et al. 1965). Cardiovascular system failure is one of the major causes of death in patients with MPS-I (Krovetz et al. 1965). "Sudden death" has been reported in patients with MPS-I, which is thought to be due to coronary artery disease or arrhythmias due to primary myocardial involvement (Krovetz et al. 1965; Yano et al. 2009). Autopsy specimens from a patient with MPS-I showed enlarged heart with markedly thickened left ventricular walls, thickened aortic and mitral valves, and endocardial fibroelastosis. Microscopic studies showed the findings of hypertrophic cardiac muscle fibers, diffuse increase in fibrous tissues, and stenosis of the major coronary arteries (Yano et al. 2009). The stenotic lesions are mainly due to thickening of the intima. Histopathologic similarity in the coronary artery lesions between the atherosclerotic changes in adults and in MPS-I has been reported (Renteria et al. 1976; Brosius and Roberts 1981).

The mechanisms which cause cardiovascular changes including coronary artery stenosis, endocardial fibroelastosis, thickened valvular lesions, and hypertrophic cardiomyopathy in MPS-I have not been well characterized. Immunohistochemical studies were conducted with the canine MPS-I models and demonstrated increased fibronectin and transforming growth factor beta-1 (TGF-β1) signaling in the vascular lesions (Lyons et al. 2011). Involvement of over expression of TGF-β1 signaling in cardiomyopathy and cardiovascular fibrosis has been reviewed (Ruiz-Ortega et al. 2007; Khan and Sheppard 2006).

Immunohistochemical studies were conducted in the cardiac specimens to evaluate transforming growth factor-beta (TGF- β) activities in the coronary arteries, endocardium, and myocardium to find out its involvement in the coronary and cardiac lesions in a patient with MPS-I (Yano et al. 2009).

Methods

This study was approved by the University of Southern California Institutional Review Board (HS-10-00375). Phosphorylated Smad2 (p-Smad2) immunofluorescent staining was performed on cardiac specimens from the patient with MPS-I previously reported (Yano et al. 2009). Formalin-fixed 5 μm sections were prepared from paraffin-embedded cardiac specimens and mounted on poly-L-lysine–coated slides. The slides were then deparaffinized in xylene and rehydrated. After incubation with primary antibody against p-Smad2 (rabbit polyclonal, Cell Signaling Technology), Cy3-conjugated donkey anti-rabbit secondary antibody (Vector Laboratories) was applied for one hour at room temperature. Sections were preserved in VECTASHELD mounting medium with DAPI (to visualize nuclei).

Results

Phosphorylated Smad2 staining in postmortem cardiac specimens from a patient with MPS-I showed increased activities in the intimal layer with myointimal proliferation as well as in the tunica adventia (Fig. 1a). Figure 1b showed significantly increased activities of phosphorylated Smad2 in the left myocardium. The age matched control showed very few phosphorylated Smad2 signals in the vascular walls as well as in the myocardium (Fig. 1c).

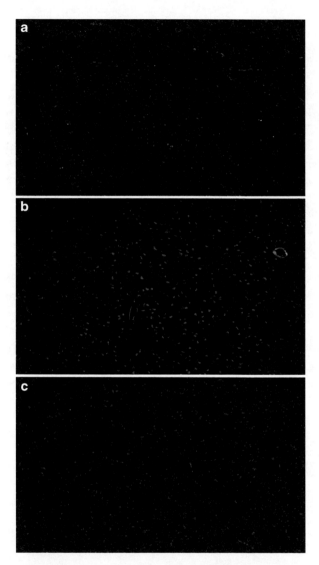

Fig. 1 Phosphoryrated Smad2 (p-Smad2) stain. (a) Patient X100, (b) Patient X 200, (c) Control X 100. (a) Extramural coronary artery section showing strong positive p-Smad2 signals (red fluorescence) in the intimal and external vascular wall. Note myointimal proliferation with prominent alpha smooth muscle signals (green fluorescence). (b) Myocardial cells showing strong p-Smad2 signals throughout the section. (c) Section showing myocardial cells and intramural coronary arteries from an age matched control with very few p-Smad2 signals compared to the patient

These findings suggest that TGF-β signal pathway is involved in coronary artery myointimal proliferation causing stenosis and myocardial hypertrophy with fibrotic changes (Fig. 1a, b). Phosphorylated Smad2 signals were increased in the thickened endocardium compared to the age matched control: 40% of cells were positive for phosphorylated Smad2 signals in the patient. Less than 5% were positive in the control (data not shown).

Discussion

There was no specific treatment available for any type of MPS until bone marrow transplantation was first reported in 1981 by Hobbs (Hobbs et al. 1981). Since enzyme replacement therapy (ERT) for MPS-I was introduced in 2003, it has been widely recognized that some clinical symptoms progress despite ERT. Mental retardation, for example, is well recognized and predicted to progress since the therapeutic enzyme can not go through the blood brain barrier. Development of cardiovascular complications including cardiomyopathy and valvular abnormalities are also well known in patients with MPS-I who have been treated with ERT (Sifuentes et al. 2007).

To study the mechanisms causing hypertrophic cardiomyopathy with fibrosis in the left ventricular wall, endocardial fibroelastosis, and the coronary artery lesions with intimal proliferation resulting in stenosis, immunohistochemical studies were performed on the pathology specimens from the patient with MPS-I to evaluate involvement of TGF-β activities. The study showed significantly increased phosphorylated Smad2 signals in the myocardial cells, in the endocardial cells, and in the coronary artery walls in the specimens from the patient with MPS-I. These findings suggest that TGF-β signal pathway is involved in these lesions (Fig. 1a, b). In mammals, TGF-β has three isoforms: TGF-β1, TGF-β2, and TGF-β3. TGF-β1 is expressed in myofibroblasts, vascular smooth muscle cells, endothelial cells, and macrophages. TGF-β1 leads to phosphorylation of Smad2 protein through the process of binding to a dimerized receptor, consisting of TGF-β1 receptor 1 and TGF-β1 receptor 2, found on the cell surface (Khan and Sheppard 2006). In the heart, it has been postulated that the effects of TGF-β1 are primarily mediated through Smad2 phosphorylation (Pokharel et al. 2002).

Although the primary cause of TGF-β hyperactivity in the cardiovascular lesions in a patient with MPS-I is unclear, increased TGF-β activity is known to up-regulate beta1,3-glucuronosyltransferase-I in fibroblasts, a rate-limiting enzyme in glycosaminoglycan synthesis, which leads to increase synthesis of glycosaminoglycans including dermatan sulfate, hyaluronic acid, and chondroitin sulfate

A and C (Venkatesan et al. 2011; Matalon and Dorfman 1968). It is also known that accumulated dermatan sulfate can activate STAT proteins which increase productions of elastin-degrading proteins, i.e., matrix metalloproteinase-12 (MMP-12) and cathepsin S. (Ma et al. 2008). Reduction of elastin in myocardial cells, endocardium, and the coronary artery may lead to proliferation of connective tissues, resulting in fibrosis in the ventricular walls, endocardium, and vascular walls (Hinek and Wilson 2000; Karnik et al. 2003).

As reported (Yano et al. 2009), ERT may not be able to prevent or improve the cardiovascular changes. It has become widely known among treating physicians that skeletal lesions and cardiac lesions, particularly valvular lesions, often do not respond to ERT. The observation of TGF-β1 hyperactivity in the cardiovascular lesions in a patient with MPS-I may lead to a new therapy to prevent the life threatening cardiovascular changes.

In summary, we showed evidence of involvement of TGF-β hyperactivity in the cardiovascular lesions including hypertrophic cardiomyopathy, endocardial fibroelastosis, and the coronary stenosis in a patient with MPS-I. Our findings may suggest that medications which inhibit TGF-β activities, such as losartan, may have beneficial effects in preventing these life threatening complications in patients with MPS-I. Immunohistochemical studies in patients with other types of MPS are also indicated since the similar pathogenesis may exist. Further studies are warranted to evaluate the effectiveness of the TGF-β inhibitors on cardiovascular lesions in patients with MPS.

Synopsis

TGF-β hyperactivity is involved in cardiovascular lesions in MPS-I.

References to Electronic Databases

MPS-I (Hurler disease) MIM 607014
Transforming growth factor-beta (TGF-β)

References

Brosius FC, Roberts WC (1981) Coronary artery disease in the Hurler syndrome: qualitative and quantitative analysis of the extent of coronary narrowing at necropsy in six children. Am J Cardiol 47 (3):649–653

Hinek A, Wilson SE (2000) Impaired elastogenesis in Hurler disease: dermatan sulfate accumulation linked to deficiency in elastin-binding protein and elastic fiber assembly. Am J Pathol 156 (3):925–938

Hobbs JR, High-Jones K, Barrett AJ et al (1981) Reversal of clinical features of Hurler's disease and biochemical improvement after treatment by bone-marrow transplantation. Lancet 2 (8249):709–712

Karnik SK, Brooke BS, Bayes-Genis A et al (2003) A critical role of elastin signaling in vascular morphogenesis and disease. Development 130(2):411–423

Khan R, Sheppard R (2006) Fibrosis in heart disease: understanding the role of transforming growth factor-β1 in cardiomyopathy, valvular disease and arrhythmia. Immunology 118:10–24

Krovetz LJ, Lorincz AE, Schiebler GL (1965) Cardiovascular manifestations of the Hurler syndrome: hemodynamic and angiocardiographic observations in 15 patients. Circulation 31:132–141

Lyons JA, Dickson P, Wall J et al (2011) Arterial pathology in canine mucopolysaccharidosis-1 and response to therapy. Lab Invest 91 (5):665–674

Ma X, Tittiger M, Knutsen RH et al (2008) Upregulation of elastase proteins results in aortic dilation in mucopolysaccharidosis I mice. Mol Genet Metab 94(3):298–304

Matalon R, Dorfman A (1968) The structure of acid mucopolysaccharides produced by hurler fibroblasts in tissue culture. Proc Natl Acad Sci USA 60(1):179–185

Pokharel S, Rasoul S, Roks AJM et al (2002) N-acetyl-Ser-Asp-Lys-Pro inhibits phosphorylation of Smad2 in cardiac fibroblasts. Hypertension 40:155–161

Renteria VG, Ferrans VJ, Roberts WC (1976) The heart in the Hurler syndrome: gross, histologic and ultrastructural observations in five necropsy cases. Am J Cardiol 38(4):487–501

Ruiz-Ortega M, Rodriguez-Vita J, Sanchez-Lopez E, Carvajal G, Egido J (2007) TGF-β signaling in vascular fibrosis. Cardiovasc Res 74:196–206

Sifuentes M, Doroshow R, Hoft R et al (2007) A follow-up study of MPS I patients treated with laronidase enzyme replacement therapy for 6 years. Mol Genet Metab 90(2):171–180

Venkatesan N, Ouzzine M, Kolb M, Netter P, Ludwig MS (2011) Increased deposition of chondroitin/dermatan sulfate glycosaminoglycan and upregulation of β1,3-glucuronosyltransferase I in pulmonary fibrosis. Am J Physiol Lung Cell Mol Physiol 300(2):L191–L203

Yano S, Moseley K, Pavlova Z (2009) Postmortem studies on a patient with mucopolysaccharidosis type I: histopathological findings after one year of enzyme replacement therapy. J Inherit Metab Dis. doi 10.1007/s10545-009-1057-4

JIMD Reports
DOI 10.1007/8904_2012_142

RESEARCH REPORT

Recommendations for Pregnancies in Patients with Crigler-Najjar Syndrome

J. H. Paul Wilson · Maarten Sinaasappel ·
Fred K. Lotgering · Janneke G. Langendonk

Received: 5 August 2011 /Revised: 24 February 2012 /Accepted: 12 March 2012 /Published online: 22 April 2012
© SSIEM and Springer-Verlag Berlin Heidelberg 2012

Abstract During pregnancy, the developing foetus in mothers with Crigler-Najjar type 1 and 2 is exposed to raised levels of unconjugated bilirubin, with the risk of neurotoxicity. We describe two pregnancies in a patient with Crigler-Najjar type 2, who was carefully monitored prior to and during pregnancy and phototherapy adjusted to maintain serum bilirubin levels below 200 µmol/l and the bilirubin/albumin molar ratio below 50%. Both pregnancies resulted in normal delivery of healthy infants who had normal neurological development. A review of all reported pregnancies in Crigler-Najjar patients and a set of recommendations are presented.

Introduction

Crigler-Najjar syndromes type 1 (OMIM # 218800) and type 2 (OMIM # 606785) are rare disorders caused by homozygous or compound heterozygous mutations in the gene UGT1A1 and characterised by markedly raised levels of unconjugated bilirubin. Before the introduction of

Communicated by: Robin Lachmann

Competing interests: None declared

J.H.P. Wilson · J.G. Langendonk (✉)
Department of Internal Medicine, Internal medicine, metabolic diseases Office D416 Erasmus MC, PO Box 2040, 3000
CA Rotterdam, The Netherlands
e-mail: j.langendonk@erasmusmc.nl

M. Sinaasappel
Department of Paediatrics, Erasmus MC, Rotterdam,
The Netherlands

F.K. Lotgering
Department of Obstetrics and Gynaecology, Radboud University Nijmegen Medical Centre, Nijmegen, The Netherlands

phototherapy, many patients died of kernicterus in the neonatal period. Patients now can expect to survive into adulthood when treated with intensive daily phototherapy or, in the case of Crigler-Najjar type 2, with phenobarbitone with or without phototherapy. There have been very few reports of the effects of maternal unconjugated hyperbilirubinaemia on the foetus. The main concern is that the unconjugated bilirubin, which can cross the placental barrier, might cause foetal damage, especially foetal kernicterus.

In the Gunn rat, which is an animal model of the Crigler-Najjar disease, the jaundiced female rat has been reported to be less fertile than the non-jaundiced one (Davis and Yeary 1979). In addition, there seems to be an increased chance of foetal abnormalities in the jaundiced Gunn female rat (Davis et al. 1983). In humans, patients with acute and chronic liver disease also tend to be less fertile and to have an increased chance of abortion, or foetal abnormality and possibly kernicterus (Furhoff 1974; Roszkowski and Pizarek-Miedzinska 1968; Cotton et al. 1981). However, conclusions based on findings in acute and chronic hepatitis or cirrhosis cannot be extrapolated to unconjugated hyperbilirubinaemia without other features of liver disease.

At the time one of our patients expressed the desire to have children, four case reports had been published on pregnancies in patients with Crigler-Najjar disease, with five successful pregnancies (Cahill and McCarthy 1989; Taylor et al 1991; Smith and Baker 1994; Ito et al 2001). Three concerned patients with Crigler-Najjar type 2, one of whom continued to take phenobarbitone throughout the pregnancy. All were delivered of healthy infants at term. The fourth patient was diagnosed with Crigler-Najjar type 1, with total blood bilirubin levels between 357 and 458 µmol/l throughout pregnancy (1 mg/dl is 17.1 µmol/l). She underwent caesarean section. The infant had a cord total

Table 1 Recommendations for the management of pregnancy in patients with Crigler-Najjar disease types 1 and 2

1. Discussion of potential problems and genetic counselling before becoming pregnant
2. Administration of folic acid prior to and during pregnancy
3. Regular assay of serum bilirubin and albumin and adjustment of phototherapy to keep the serum bilirubin < 200 μmol/l and the bilirubin/albumin molar ratio < 50%
4. If the patient is on Phenobarbital, reduce dosage to 50–60 mg/day
5. Prenatal care and delivery in a tertiary care centre, with facilities for rapid bilirubin and albumin assays and phototherapy for mother and infant
6. Avoidance of drugs that increase unbound, unconjugated bilirubin (Robertson et al. 1991; Strauss et al. 2006)
7. Neurological testing of the child after birth and during early childhood

bilirubin of 410 μmol/l, unconjugated 217 μmol/l. This high percentage of direct reacting bilirubin is unexpected, and as the total bilirubin level of the mother and the cord blood are so similar, it could represent an artefact. Initially there were no neurological abnormalities and the infant appeared to develop normally but he was found to have some spasticity at 9 months of age (Taylor et al. 1991). We drew up a set of recommendations (Table 1), which we presented to and discussed with the Crigler-Najjar patients and their families, emphasising the limited evidence on the effects of raised maternal bilirubin levels on the developing foetus.

We have managed two successful pregnancies in a Crigler-Najjar type 2 patient, according to our guidelines. Both children are developing normally, both physically and mentally. We reviewed all reported pregnancies in Crigler-Najjar patients discovered by searching PubMed and Embase using the terms Crigler-Najjar and pregnancy.

Case Report

A woman with Crigler-Najjar disease type 2, who had been followed since childhood in our hospital, had maintained her bilirubin levels between 200 and 250 μmol/l with 1–2 h of phototherapy (10 lamps of 100 W) per day. She did not use Phenobarbital because of the side effects. She was compound heterozygote for the mutations L15R and S191F in the UDT1A1 gene.

At the age of 27 years, she became pregnant. At 9 weeks gestation her serum bilirubin was 234 μmol/l and albumin 43 g/l. Total bilirubin and direct bilirubin were determined using commercially available diazo-based colorimetric assays on a Roche Modular Analytics P 800-Module (Roche Diagnostics Nederland, Almere, The Netherlands). As during the course of her disease less than5% of the total bilirubin was direct reacting, in the later phases of the pregnancy, only total bilirubin was measured. The phototherapy was increased to 3 h/day resulting in bilirubin levels between 165 and 205 μmol/l. Her serum albumin

dropped to 34 g/l. Delivery was forceps-assisted after induction because of mild proteinuria at 39 weeks. At delivery, her bilirubin was 239 μmol/l. The healthy boy weighed 3440 g and had an Apgar score of 9 at 5 min. Cord blood bilirubin levels were the same as the mother's. The mother's bilirubin rose after delivery to 278 μmol/l and her albumin dropped to 30 g/l (Fig. 1). Directly after birth, her son's bilirubin was 177 μmol/l with albumin 35 g/l; phototherapy was given to the child for 3 days, bilirubin dropped to 10 μmol/l with unchanged albumin.

Two years later, she gave birth to a healthy girl at 39 weeks, weighing 3,150 g with an Apgar score of 10 at 5 min. Bilirubin and albumin levels were similar to the first pregnancy. Both children have developed normally without hyperbilirubinaemia or neurological abnormalities during a follow up of 11 and 9 years, respectively.

Discussion

Since the second pregnancy of our patient, there have been five more case reports on the outcome of pregnancy in patients with Crigler-Najjar type 1 and 2, giving a total of four Crigler-Najjar type 1 patients and eight Crigler-Najjar type 2, and 17 deliveries. The reports are summarised in Table 2. As can be seen, no neurological damage has been detected in the offspring, with one exception. The one exception was a Crigler-Najjar type 1 patient reported by Taylor et al. (1991) who had previously had three first trimester therapeutic abortions performed for social reasons and a prior miscarriage at which time she had a bilirubin level of 460 μmol/l. This patient is not reported to have been treated for the hyperbilirubinaemia. In all the other patients it has been possible to reduce maternal unconjugated bilirubin levels by increasing the intensity or duration of the phototherapy, in two instances in combination with fortnightly albumin infusions to increase bilirubin binding capacity or and in some Crigler-Najjar type 2 patients by continuing low dose phenobarbital treatment. As expected, maternal bilirubin levels were higher in Crigler-Najjar type 1, who also required much longer daily exposure to blue light – from 12 to 16 h.

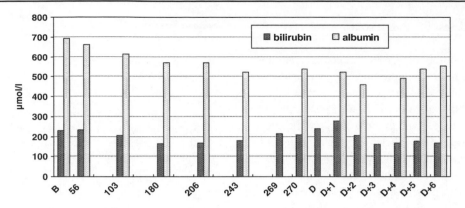

Fig. 1 Maternal serum bilirubin and albumin in μmol/l during pregnancy and after delivery. B = mean values in the 2-year period prior to pregnancy; numbers = duration of pregnancy in days; D = delivery; D + number = day after delivery

Table 2 Reports of Crigler-Najjar and pregnancy in the literature

Reference	CN type	Maternal age (years)	Maternal bilirubin during pregnancy (μmol/l)/	Maternal bilirubin at delivery (μmol/l)/	Cord blood or 1st day bilirubin (μmol/l)	Maternal treatment	Mode of delivery	Newborn treatment	Foetal outcome
Taylor 1991	NA/ 1	21	363–458	420	410 CB	None	CS	PT	Quadriplegic at 18 months
Gajdos 2006	1	28	230–280 (400 before PT)	242	222 CB	PT albumin	CS	PT	Healthy
Hannam 2009	1	26	200–380 (526 before PT)	317	323 D1	PT	CS	PT + Phb	Healthy
Hannam 2009	1	22	399–510	480	420 CB	PT albumin	Spontaneous	PT + exchange transfusion	Healthy
Cahill 1989	2	22	178–183	104	121 D1	Phb	Induced	PT	Healthy
Smith 1994	2	24	91–164	150	130 CB	None	Spontaneous	None	Healthy
Ito 2001	2	< 34	55–116			PT + Phb	CS	None	Healthy
			75–173	65–94	87 CB	PT + Phb	CS	None	Healthy
Holstein 2005 Holstein 2011	2	36	72–94	NA	NA	PT + Phb	CS	None	Healthy
		38	72–94	NA	NA	Phb	CS	None	Healthy
		41	72–94	NA	NA	Phb	CS	None	None
Saxena 2005	2	24	NA	139	174 D1	None	Spontaneous	None	Healthy
		25	130			None	Spontaneous	Blood transfusion, PT + Phb	Healthy
Passuello 2009	2	37	68–154	67	94 D1	Phb	Spontaneous	None	Healthy
Arora 2009	2	26	137–144	144	227 D1	None	Spontaneous	PT	Healthy
Current case	2	27	165–205	239	187 D1	PT	Induced	PT	Healthy
		29	168–256	195	146 D1	PT	Spontaneous	None	Healthy

Features of pregnancies in patients with Crigler-Najjar reported in the literature, *PT* phototherapy, *Phb* phenobarbital, *CS* caesarean sections, *NA* not available, *CB* cord blood, *D1* blood taken on first day after delivery. Bilirubin levels reported in mg/dl were converted to μmol/l by multiplying by 17.1. Note that in the report by Saxena (2005) cord blood levels showed a surprisingly high level of conjugated bilirubin, which might be an artefact

In general, low doses of phenobarbital have been given (25–100 mg/day).

Unconjugated bilirubin can cross the placenta, and this is indeed the normal route for excretion during pregnancy as the foetal liver is unable to conjugate or excrete bilirubin (Macias et al. 2009; Gajdos et al. 2006; McDonagh 2007; 2010). It is conceivable that phototherapy of the mother might reduce exposure of the foetal brain to a greater extent than reflected by serum bilirubin levels as the photoisomers of bilirubin are more water soluble and do not cross the blood–brain barrier or presumably the placenta to the same extent as unconjugated bilirubin (Mreihil 2010). At birth, foetal unconjugated bilirubin levels are in most reports roughly equivalent to those found in their mothers with Crigler-Najjar disease; however, phototherapy is usually stopped several hours before delivery, and circulating levels of photoisomers would be expected to be decreasing. However, serum levels were lower than the mother's in both infants reported here, which is compatible with a reduced clearance of photoisomers by the placenta.

At present it is impossible to state what concentration of unconjugated bilirubin is non-neurotoxic for the developing foetus, but it seems reasonable to try to keep the maternal (and therefore the foetal) unconjugated bilirubin concentrations below 200 μmol/l and the molar ratio between unconjugated bilirubin and serum albumin below 50% (Gajdos et al. 2006; Strauss et al. 2006). Current national guidelines state that a bilirubin over 170 μmol/l is an indication for phototherapy in term infants without risk factors for kernicterus and 70 μmol/l in high-risk infants.

In most patients with Crigler-Najjar who attended antenatal clinics, it has been possible to achieve these levels by increasing phototherapy, and, if needed, by infusions of human albumin.

Admission to hospital to facilitate monitoring and treatment of both mother and infant is necessary – as can be seen by the frequent need for phototherapy of the neonate. Careful planning of the admission, including use of a written protocol and frequent visits by the metabolic diseases team to monitor and educate the medical and nursing staff of the obstetrics unit, is essential.

Conclusions

Regular monitoring and adjustment of treatment to keep bilirubin levels below 200 μmol/l and bilirubin/albumin molar ratios below 50% during pregnancy in patients with Crigler-Najjar syndrome can result in good foetal outcome.

Conflict of Interest Statement

None.

References

Arora N, Choudhary S (2009) Pregnancy with Crigler-Najjar syndrome type II. J Obstet Gynaecol 29:242–244

Cahill DJ, McCarthy CF (1989) Pregnancy and the Crigler-Najjar syndrome. J Obstet Gynecol 9:213

Cotton BD, Brock BJ, Schifrin BS (1981) Cirrhosis and fetal hyperbilirubinemia. Obstet Gynecol 57(6 suppl):25–27S

Davis DR, Yeary RA (1979) Impaired fertility in the jaundiced female (Gunn) rat. Lab Anim Sci 29:739–743

Davis DR, Yeary RA, Lee K (1983) Improved embryonic survival in the jaundiced female rat fed activated charcoal. Pediatr Pharmacol 3:79–85

Furhoff AK (1974) Fate of children born to women with jaundice in pregnancy. Arch Gynäk 217:165–172

Gajdos V, Petit F, Trioche P et al (2006) Successful pregnancy in a Crigler-Najjar type I patient treated by phototherapy and semimonthly albumin infusions. Gastroenterology 131:921–924

Hannam S, Moriaty P, O'Reilly H et al (2009) Normal neurological outcome in two infants treated with exchange transfusions born to mothers with Crigler-Najjar Type 1 disorder. Eur J Pediatr 168:427–429

Holstein A, Plaschke A, Lohse P, Egberts EH (2005) Successful photo-and phenobarbital therapy during pregnancy in a woman with Crigler-Najjar syndrome type II. Scand J Gastroenterol 40:1124–1126

Holstein A, Bryan CS (2011) Three consecutive pregnancies in a woman with Crigler-Najjar syndrome type II with good maternal and neonatal outcomes. Dig Liver Dis 43:170

Ito T, Katagiri C, Ikeno S, Takahashi H, Nagata N, Terakawa N (2001) Phenobarbital following phototherapy for Crigler-Najjar syndrome type II with good fetal outcome: a case report. J Obstet Gynaecol Res 27:33–35

Macias RI, Marin JJ, Serrano MA (2009) Excretion of biliary compounds during intrauterine life. World J Gastroenterol 15:817–828

McDonagh AF (2007) Movement of bilirubin and bilirubin conjugates across the placenta. Pediatrics 119:1032–1033

McDonagh AF (2010) Controversies in bilirubin biochemistry and their clinical relevance. Semin Fetal Neonatal Med 15:141–147

Mreihil K, McDonagh AF, Nakstad B, Hansen TW (2010) Early isomerization of bilirubin in phototherapy of neonatal jaundice. Pediatr Res 67:656–659

Passuello V, Puhl AG, Wirth S et al (2009) Pregnancy outcome in maternal Crigler-Najjar syndrome type II: a case report and systematic review of the literature. Fetal Diagn Ther 26:121–126

Robertson A, Karp W, Brodersen R (1991) Bilirubin displacing effect of drugs used in neonatology. Acta Paediatr Scand 80:1119–1127

Roszkowski I, Pizarek-Miedzinska D (1968) Jaundice in pregnancy. II. Clinical course of pregnancy and delivery and condition of neonate. Am J Obstet Gynecol 101:500–503

Saxena P, Arora R, Minocha B (2005) Crigler-Najjar syndrome with pregnancy. J Obstet Gynecol India 55:270–271

Smith JF Jr, Baker JM (1994) Crigler-Najjar disease in pregnancy. Obstet Gynecol 84:670–672

Strauss KA, Robinson DI, Vreman HK, Puffenberger EG, Hart G, Morton DH (2006) Management of hyperbilirubinemia and prevention of kernicterus in 20 patients with Crigler-Najjar disease. Eur J Pediatr 165(5):306–319

Taylor WG, Walkinshaw SA, Farquharson RG, Fisken RA, Gilmore IT (1991) Pregnancy in Crigler-Najjar syndrome. Br J Obstet Gynaecol 98:1290–1291

JIMD Reports
DOI 10.1007/8904_2012_143

CASE REPORT

Autism Spectrum Disorder in a Child with Propionic Acidemia

M. Al-Owain · N. Kaya · H. Al-Shamrani ·
A. Al-Bakheet · A. Qari · S. Al-Muaigl ·
M. Ghaziuddin

Received: 28 September 2011 / Revised: 01 March 2012 / Accepted: 12 March 2012 / Published online: 31 March 2012
© SSIEM and Springer-Verlag Berlin Heidelberg 2012

Abstract Autism is a neurodevelopmental disorder characterized by a combination of reciprocal social deficits, communication impairment, and rigid ritualistic interests. While autism does not have an identifying cause in most of the cases, it is associated with known medical conditions in at least 10% of cases. Although uncommon, cases of autism have also been reported in association with metabolic disorders. In this brief report, we describe the occurrence of autism in a 7-year-old girl with propionic acidemia (PA), a common form of organic aciduria resulting from the deficiency of propionyl-CoA carboxylase and characterized by frequent and potentially lethal episodes of metabolic acidosis often accompanied by hyperammonemia. It is particularly common in countries with high rates of consanguinity. Early diagnosis of autism in patients with metabolic disorders is important since autistic features are sometimes the most disruptive of all the child's problems. This facilitates providing the needed behavioral services not otherwise available for children with metabolic disorders.

Communicated by: Ivo Barić

Competing interests: None declared

M. Al-Owain (✉) · A. Qari
Department of Medical Genetics, King Faisal Specialist Hospital
and Research Centre, PO Box 3354, Riyadh 11211, Saudi Arabia
e-mail: alowain@kfshrc.edu.sa

N. Kaya · A. Al-Bakheet
Department of Genetics, King Faisal Specialist Hospital and
Research Centre, Riyadh, Saudi Arabia

H. Al-Shamrani
Department of Pediatrics, King Faisal Specialist Hospital and
Research Centre, Riyadh, Saudi Arabia

S. Al-Muaigl
Department of Nutrition, King Faisal Specialist Hospital and
Research Centre, Riyadh, Saudi Arabia

M. Ghaziuddin
Department of Psychiatry, King Faisal Specialist Hospital and
Research Centre, Riyadh, Saudi Arabia

M. Al-Owain
College of Medicine, Alfaisal University, Riyadh, Saudi Arabia

Abbreviations

PA   Propionic acidaemia
PPA   Propionic acid
ASD   Autism spectrum disorder

Introduction

Autism is a neurodevelopmental disorder characterized by a distinct combination of social and communication deficits with rigid restricted interests (Centers for Disease Control and Prevention 2007). It is classified as the main category in a group of disorders, called autism spectrum (or pervasive developmental) disorders (ASD), all of which are characterized by similar deficits (Schaefer and Lutz 2006). Although generally regarded as a disorder with strong genetic underpinnings, it is associated with known medical conditions in a subset of cases (Folstein and Rosen-Sheidley 2001; Pickler and Elias 2009). These medical conditions include chromosomal abnormalities, such as, fragile X syndrome (Clifford et al. 2007); neurological disorders, such as, tuberous sclerosis (Curatolo et al. 2010); and a range of inborn errors of metabolism. Among the latter, the most prominent include phenylketonuria, disorders of purine and pyramidine metabolism, glucose transport disorders, and mitochondrial disorders (Kayser 2008; Schaefer and Lutz 2006; Zecavati and Spence 2009). To our knowledge, autism has not been reported in association with propionic acidemia (PA), an autosomal recessive metabolic disorder characterized by frequent and potentially lethal episodes of ketoacidosis and hyperammonemia. In this report, we describe a case of PA with autism.

Case Report

This Saudi girl, a product of uneventful pregnancy with normal birth growth parameters, was diagnosed with PA since birth based on tandem mass spectrometry and urine gas chromatography–mass spectrometry. *PCCA* gene sequencing revealed a homozygous G117D mutation. Family history was notable for parental consanguinity, but no history of neurodevelopmental disorders. At the age of 6 weeks, she presented with hyperammonemia (331 μmol/l) without metabolic acidosis. Subsequently, she had frequent episodes of metabolic acidosis, hyperammonemia (100–270 μmol/l) and recurrent pancreatitis (six attacks). She was treated with carnitine and sodium benzoate (250–400 mg/kg/day), which was changed later to carglumic acid (N-carbamylglutamate, 50 mg/kg/day) due to recurrent hyperammonemic episodes. Formal developmental assessment at the age of 8 years revealed that her cognitive and language skills coincided with the level of 21–24 months of age, while the gross and fine motor skills were at the level of 21–24 and 24 months of age, respectively. Her personal/social skills were at the level of 24–30 months of age. Nutritionally, she was on Propimex®-2 (66.5 g), polycose (80 g), whole milk (378 ml), and solid food providing total protein of 1.4 g/kg/day and natural protein of 0.68 g/kg/day. On physical exam, she was conscious with weight of 30 kg (90%), height of 124 cm (75%), and occipitofrontal circumference of 51 cm (2–50%). Chest, heart, and abdominal examination was unremarkable. Central nervous system examination revealed only mild hypotonia. Evaluation by electroencephalography and computed tomography of the brain did not reveal any abnormalities. The parents refused the brain MRI evaluation. Repeatedly, her propionylcarnitine was 106–114 (normal: <10 μmol/l) with a C3/C2 ratio of 4–5.89 (normal: 0–0.4). The free carnitine was 100–287 (normal: 6–72 μmol/l). *FMR1* gene testing revealed that she had normal CGG repeats of 22 and 29. Array-comparative genomic hybridization and *MECP2* gene sequencing were normal.

Psychiatric Evaluation

The patient was referred to the child psychiatrist because of behavioral problems consisting mainly of her difficulty in interacting with other children of her age, poor language skills, and resistance to follow commands. According to her parents, behavioral symptoms were noted before she was

3 years old and were initially attributed to her global intellectual disability. She used to spend several hours during the day watching cartoons on the DVD player and on the television. She would rewind the tape and watch her favorite scenes repeatedly, getting irritable and angry if stopped. At times, this would lead to severe temper tantrums, which would end only if her parents let her resume watching the cartoons again. She also liked spinning colored pens, holding them in a certain manner, and flicking them repeatedly. Other repetitive behaviors consisted of walking in circles; flapping her hands, especially when excited; switching lights on and off repeatedly; and occasionally smelling her mother's hair. Her imaginative play was minimal. With other children her of age, she would tend to isolate herself and not show any interest in joining them. Her conversation was limited to a few words, such as, bed-sheet or food items. However, she tended to repeat lines from her favorite cartoon shows, such as, *Mickey Mouse* and *Tom and Jerry*. She did not usually look at people when speaking to them nor would she point out to objects at a distance. Her facial expression, too, was limited. During the interview, she was well groomed and appeared her stated age. She was not aggressive or violent. However, she did not appear to be aware of personal boundaries; for example, she would occasionally stretch her hand across the desk and try to grab the pen from the examiner's hand without bothering to look at him. At times, she clapped her hands, and tried to reach behind her mother's head-cover to touch her (mother's) hair and smell it, while at other times, she would walk around in circles around the examiner's desk, flicking a pen, and screaming if stopped from doing so. When parents were interviewed about her behavior based on the Social Communication Questionnaire (SCQ) (Rutter et al. 2003), she received a score of 19, which is above the cutoff for autism.

Our patient in this report showed the typical triad of autistic symptoms – reciprocal social deficits, communication impairment, and restricted repetitive interests – from early childhood. Structured informant-based interviews for autism were conducted because of the lack of such instruments in Arabic. Therefore, based on the history and the clinical examination, and supplemented by scores on the rating scale, a diagnosis of autistic disorder as defined by the DSM IV (American Psychiatric Association 2000) was made by the child's psychiatrist.

Discussion

This case report describes the occurrence of autism in a child with propionic acidemia (PA). While a chance occurrence of the two syndromes cannot be ruled out, it is possible that PA may have played a role in the emergence of autistic features in this patient because acute and chronic abnormalities of brain function are well-documented in PA (de Baulny et al. 2005; Schreiber et al. 2012). Alternatively, it is possible that the patient was already vulnerable to autism due to some other reason and that PA acted only as a trigger, a view consistent with the so-called two-hit hypothesis of the disorder (Ghaziuddin 2000). Noteworthy, urea cycle disorders, characterized by hyperammonemia, may present with confusion, bizarre behavior, and autistic-like symptoms (Gorker and Tuzun 2005; Sedel et al. 2007). It is possible that the recurrent bouts of hyperammonemia played a role in the development of the autistic symptoms in our patient. In the consensus conference about diagnosis and management of PA hosted in Washington, D.C. in January 2011, there was no reported association among the neurological sequalae of the disease between PA and autism (Schreiber et al. 2012; Sutton et al. 2012). This could be related to the fact that children with severe-profound intellectual disability may have impaired social interactions and the diagnosis of ASD might be challenging.

Recent reports have suggested that propionic acid (PPA) may be used to induce an animal model of autism (MacFabe et al. 2007, 2008). PPA, a short-chain fatty acid, is an intermediate of cellular metabolic pathways. As a weak acid, PPA exists in both aqueous and lipid soluble forms; it crosses the blood-brain barrier both passively and actively through specific monocarboxylate transporters (Thomas et al. 2010). Intracerebroventricular injection of PPA in adult rats produced behavioral, biochemical, electrophysiological, and neuropathological effects similar to those observed in autism (MacFabe et al. 2007, 2008; Shultz et al. 2009, 2008). Following intracerebroventricular infusions of PPA, there was an evidence of a relationship between changes in brain lipid profiles and the occurrence of autistic behaviors (Thomas et al. 2010). Collectively, the effects of PPA in rats included reversible repetitive dystonic behaviors, hyperactivity, turning behavior, retropulsion, caudate spiking, and the progressive development of limbic kindled seizures (MacFabe et al. 2007). Similarly, systemic administration of PPA has been shown to cause social deficits, anxiety-like and hypoactivity behavior in juvenile rats (Shams et al. 2009).

There is a strong correlation of gastrointestinal symptoms with autism severity in children with ASDs (Adams et al. 2011). Within the group of children with autism and gastrointestinal disturbances, unique impairments of carbohydrate digestion and transport, and mucosal dysbiosis were found (Williams et al. 2011). This underlies the hypothesis that there exists a relationship between increased production of enteric short chain fatty acids, including PPA from the altered carbohydrate fermentation and the long-term development of ASD-related behavioral changes (Ossenkopp et al. 2012).

Our patient's symptoms could not be explained on the basis of global intellectual disability alone. Compared to other intellectually disabled children, her symptoms had a unique and specific cluster marked by a distinct combination of reciprocal social deficits (as shown by her desire to be alone, difficulty in interacting with other children of the same cognitive level, inability to read the emotions and feelings of others, etc.), communication impairment (poor eye-contact and gesture, presence of echolalia, language delay etc.), and restricted range of interests/repetitive behaviors (watching the same cartoon shows, stereotyped body movements, sensory abnormalities such as touching hair, etc.) that strongly supported an additional diagnosis of autism. On a reliable and valid screening measure of autism, the Social and Communication Questionnaire (SCQ) (Rutter et al. 2003), she scored above the cutoff for that disorder. A semi-structured interview for autism, such as, the Autism Diagnostic Interview-Revised was not used because such instruments have not been translated into Arabic.

The association of PA with autism has both clinical and research implications. From a clinical standpoint, children with PA, especially in countries where its rates are high, should be routinely screened for autism because early diagnosis and intervention lead to a better outcome (Howlin et al. 2004). Conversely, children with autism should receive a thorough genetics evaluation using current technology (Folstein and Rosen-Sheidley 2001; Pickler and Elias 2009). Although screening for metabolic disorders is not recommended in all children with autism (Filipek et al. 1999), it may be necessary in selected patients, especially in countries with high rates of recessive disorders, such as Saudi Arabia. For example, affected children with PA are known to suffer from intellectual disability, and, it is possible that at least some of them suffer from autism. Pediatricians, community physicians, and medical geneticists should, therefore, be trained to suspect and screen for the presence of autism. Systematic studies should also be undertaken to examine the association of PA and autism in large samples.

Acknowledgments We are grateful to the patient and her family for participation in this study. In addition, we thank Dr. Derrick MacFabe and Dr. V. Reid Sutton for their scientific contributions.

Synopsis

First report of Autism in propionic acidemia.

Competing Interests

None declared.

References to Electronic Databases

Propionic acidaemia: OMIM 606054.

Propionyl-CoA carboxylase: EC 6.4.1.3.

References

Adams JB, Johansen LJ, Powell LD et al (2011) Gastrointestinal flora and gastrointestinal status in children with autism–comparisons to typical children and correlation with autism severity. BMC Gastroenterol 11:22

American Psychiatric Association (2000) Diagnostic and statistical manual of mental disorders, Fourth edition, text revisionth edn. APA, Washington, DC

Centers for Disease Control and Prevention (2007) Autism spectrum disorders overview. http://www.cdc.gov/ncbddd/autism/overview.htm. Retrieved 03 April 2009

Clifford S, Dissanayake C, Bui QM et al (2007) Autism spectrum phenotype in males and females with fragile X full mutation and premutation. J Autism Dev Disord 37:738–747

Curatolo P, Napolioni V, Moavero R (2010) Autism spectrum disorders in tuberous sclerosis: pathogenetic pathways and implications for treatment. J Child Neurol 25:873–880

de Baulny HO, Benoist JF, Rigal O et al (2005) Methylmalonic and propionic acidaemias: management and outcome. J Inherit Metab Dis 28:415–423

Filipek PA, Accardo PJ, Baranek GT et al (1999) The screening and diagnosis of autistic spectrum disorders. J Autism Dev Disord 29:439–484

Folstein SE, Rosen-Sheidley B (2001) Genetics of autism: complex aetiology for a heterogeneous disorder. Nat Rev Genet 2:943–955

Ghaziuddin M (2000) Autism in Down's syndrome: a family history study. J Intellect Disabil Res 44(Pt 5):562–566

Gorker I, Tuzun U (2005) Autistic-like findings associated with a urea cycle disorder in a 4-year-old girl. J Psychiatry Neurosci 30:133–135

Howlin P, Goode S, Hutton J et al (2004) Adult outcome for children with autism. J Child Psychol Psychiatry 45:212–229

Kayser MA (2008) Inherited metabolic diseases in neurodevelopmental and neurobehavioral disorders. Semin Pediatr Neurol 15:127–131

MacFabe DF, Cain DP, Rodriguez-Capote K et al (2007) Neurobiological effects of intraventricular propionic acid in rats: possible role of short chain fatty acids on the pathogenesis and characteristics of autism spectrum disorders. Behav Brain Res 176:149–169

MacFabe DF, Cain NE, Boon F et al (2008) Effects of the enteric bacterial metabolic product propionic acid on object-directed behavior, social behavior, cognition, and neuroinflammation in adolescent rats: Relevance to autism spectrum disorder. Behav Brain Res 217:47–54

Ossenkopp KP, Foley KA, Gibson J et al (2012) Systemic treatment with the enteric bacterial fermentation product, propionic acid, produces both conditioned taste avoidance and conditioned place avoidance in rats. Behav Brain Res 227:134–141

Pickler L, Elias E (2009) Genetic evaluation of the child with an autism spectrum disorder. Pediatr Ann 38:26–29

Rutter M, Berument S, Lord C, Pickles A (2003) Social and Communication Questionnaire (SCQ), lifetime version. Western Psychological Services, Los Angeles

Schaefer GB, Lutz RE (2006) Diagnostic yield in the clinical genetic evaluation of autism spectrum disorders. Genet Med 8:549–556

Schreiber J, Chapman KA, Summar ML et al (2012) Neurologic considerations in propionic acidemia. Mol Genet Metab 105:10–15

Sedel F, Baumann N, Turpin JC et al (2007) Psychiatric manifestations revealing inborn errors of metabolism in adolescents and adults. J Inherit Metab Dis 30:631–641

Shams S, Kavaliers M, Foley KA et al (2009) Social deficits, anxiety-like behavior and hypoactivity following systemic administration of propionic acid in juvenile rats. Abstract [#436.6]. Society for Neuroscience, Chicago

Shultz SR, MacFabe DF, Ossenkopp KP et al (2008) Intracerebroventricular injection of propionic acid, an enteric bacterial metabolic end-product, impairs social behavior in the rat: implications for an animal model of autism. Neuropharmacology 54:901–911

Shultz SR, Macfabe DF, Martin S et al (2009) Intracerebroventricular injections of the enteric bacterial metabolic product propionic acid impair cognition and sensorimotor ability in the Long-Evans rat: further development of a rodent model of autism. Behav Brain Res 200:33–41

Sutton VR, Chapman KA, Gropman AL et al (2012) Chronic management and health supervision of individuals with propionic acidemia. Mol Genet Metab 105:26–33

Thomas RH, Foley KA, Mepham JR et al (2010) Altered brain phospholipid and acylcarnitine profiles in propionic acid infused rodents: further development of a potential model of autism spectrum disorders. J Neurochem 113:515–529

Williams BL, Hornig M, Buie T et al (2011) Impaired carbohydrate digestion and transport and mucosal dysbiosis in the intestines of children with autism and gastrointestinal disturbances. PLoS One 6:e24585

Zecavati N, Spence SJ (2009) Neurometabolic disorders and dysfunction in autism spectrum disorders. Curr Neurol Neurosci Rep 9:129–136

JIMD Reports
DOI 10.1007/8904_2012_144

Urinary Neopterin and Phenylalanine Loading Test as Tools for the Biochemical Diagnosis of Segawa Disease

Vincenzo Leuzzi · Claudia Carducci · Flavia Chiarotti · Daniela D'Agnano ·
Maria Teresa Giannini · Italo Antonozzi · Carla Carducci

Received: 4 January 2012 / Revised: 01 March 2012 / Accepted: 14 March 2012 / Published online: 18 April 2012
© SSIEM and Springer-Verlag Berlin Heidelberg 2012

Abstract *Background.* The diagnosis of autosomal dominant GTP-cyclohydrolase deficiency relies on the examination of the *GCH1* gene and/or pterins and neurotransmitters in CSF. The aim of the study was to assess the diagnostic value, if any, of pterins in urine and blood phenylalanine (Phe) and tyrosine (Tyr) under oral Phe loading test.

Methods. We report on two new pedigrees with four symptomatic and four asymptomatic carriers whose pattern of urinary pterins and blood Phe/Tyr ratio under oral Phe loading pointed to GTP-cyclohydrolase deficiency. The study was then extended to 3 further patients and 90 controls. The diagnostic specificity and sensitivity of these metabolic markers were analysed by backwards logistic analysis.

Results. Two genetic alterations segregated alternatively in Family 1 (c.631-632 del AT and c.671A > G), while exon 1 deletion was transmitted along three generations in Family 2. Neopterin and biopterin concentrations in urine clustered differently in controls under and over the age of 15. Therefore patients and controls were sub grouped according to this age. Neopterin was significantly reduced in GCH1 mutated subjects younger than 15, and both neopterin and biopterin in those older than 15. Moreover, the Phe/Tyr ratios at the second and third hour were both significantly higher in patients than in controls. Backwards logistic regression demonstrated the high diagnostic sensitivity and specificity of combined values of neopterin concentration and Phe/Tyr ratio at the second hour.

Conclusions. Pterins in urine and Phe loading test are non-invasive and reliable tools for the biochemical diagnosis of GTP-cyclohydrolase deficiency.

Communicated by: Nenad Blau

Competing interests: None declared

V. Leuzzi (✉)
Department of Pediatrics, Child Neurology and Psychiatry,
Sapienza Università di Roma, Via dei Sabelli 108,
00185 Roma, Italy
e-mail: vincenzo.leuzzi@uniroma1.it

C. Carducci · I. Antonozzi · C. Carducci
Department of Experimental Medicine, Sapienza Università di
Roma, Rome, Italy

C. Carducci
Department of Molecular Medicine, Sapienza Università di Roma,
Rome, Italy

F. Chiarotti
Department of Cell Biology and Neuroscience, Istituto Superiore
di Sanità, Rome, Italy

D. D'Agnano · M.T. Giannini
Department of Pediatrics, Child Neurology and Psychiatry,
Sapienza Università di Roma, Roma, Italy

Introduction

Autosomal dominant (AD) GTP-cyclohydrolase (GTP-CH EC 3.5.4.16) deficiency (DYT5a; MIM # 600225) presents with a wide spectrum of Dopa-responsive movement disorders (DRD) and a variable and sex-conditioned penetrance (Segawa et al. 2003; Segawa 2009). *GCH1* gene alterations are found in about 60% of patients (Segawa 2009), while in almost all cases neopterin and biopterin are low in CSF (Fink et al. 1988; Furukawa et al. 1993; LeWitt et al. 1986).

Pterins in urine, even though rarely assessed, are generally considered normal (Furukawa et al. 1998). We report on two new families with genetically confirmed DYT5a whose pattern of urinary pterins coupled with phenylalanine (Phe)/ tyrosine (Tyr) ratio under oral Phe loading pointed to GTP-GH deficiency. On this perspective, the biochemical hallmarks of the disease were reviewed and the potential diagnostic value of Phe loading test and pterins in urine was explored.

Patients

Family 1. The proband, a 12-year-old girl, was born after a normal pregnancy and delivery from Italian unrelated parents. Mother and a 9-year-old brother were normal. Her 36-year-old father had been suffering since adolescence from gait fatigability and foot rigidity that increased in the evening period. The patient's early psychomotor development was considered normal. At the age of 3 years, she suffered from atonic fits and was treated with antiepileptic medicaments. When 8 years old, she was again examined because of learning difficulties that were ascribed to a mild mental retardation. Starting from the age of 9 years, she complained of fatigability and instable gait emerging during the evening period. On examination, at the age of 11, she exhibited generalised choreoathetosis, which was exacerbated by exercise and in the late afternoon and evening when she was no longer capable of walking unsupported. She was mildly mentally retarded (WISC-R IQ 51), depressed and anxious. On neurological examination her father showed a mild foot dystonia with diurnal fluctuation.

Molecular analysis of the *GCH1* gene (see below for the methods) disclosed a frameshift mutation in exon 6 (c.631-632 del AT, p. M211fs) in the girl and in her father (case 1a and b, respectively, Table 1). Unexpectedly, a different point mutation was found in the patient's mother and brother (c.671A>G, p.K224R) (case 1c and 1d, Table 1).

L-Dopa/Carbidopa treatment (up to 8/2 mg/kg bw/day) resulted in a dramatic improvement in the father and in the proband. At the same time her WISC-R IQ improved to 63.

Family 2. A 7.5-year-old girl of Swedish and Italian descent presented with a progressive long-lasting severe gait disorder with diurnal fluctuation. She was the first offspring of two unrelated parents. The 32-year-old mother had been complaining during childhood of remarkable fatigability that disappeared in the following years. The proband was born after a normal pregnancy and delivery. She was able to walk unsupported at the age of 13 months when the parents first noticed fatigability and protracted uncertainty in the gait. At the age of 3, a mild paraparesis was diagnosed. During the following few years paraparesis worsened and a diurnal fluctuation of the symptoms became manifest. On examination, at the age of 7, she showed normal mental development (WISC-III IQ 115), generalised hypo- and bradykinesia, dystonic paraparetic gait with extra rotation of the hip and feet drop in the propulsive phase of gait and masked face with rigid fixed posture of head and neck. Exon 1 deletion in the *GCH1* gene was detected in the proband, her brother, mother and unaffected 58-year-old maternal grandmother of

Swedish origin (respectively cases 2a-2d in Table 1) (see below for the methods).

The only neurological sign detected in the mother was an exaggerated dorsal flexion of the big toe in the oscillatory phase of the gait.

Under L-Dopa/Carbidopa treatment (2/0.5 mg/kg bw/ day), the girl experienced a marked improvement of both hypo-bradykinesia and lower limb dystonia.

Methods

Genomic analysis included the following: exon and intron–exon boundaries sequencing (Bandmann et al. 1996), screening for intra-gene deletions or duplications of the alleles negative to sequencing analysis (Zirn et al. 2008) (SALSA MLPA KIT, MRC-Holland, Amsterdam, The Netherlands) (Schouten et al. 2002) and retesting of the patients positive to MLPA analysis with Real-Time PCR (SYBR Green dye chemistries).

Informed consent was obtained from all subjects examined and (if minor) their parents.

5-Hydroxyindoleacetic acid [5-HIAA], homovanillic acid [HVA], and 3-methoxy-4-hydroxyphnylglycol were assessed in CSF by high-performance liquid chromatography with electrochemical detection. Neopterin and biopterin were determined in the first morning urine sample and CSF according to the method published elsewhere (Antonozzi et al. 1988). Phe and Tyr were measured by ESI-MS/MS (Chace et al. 1993) in dried blood spots at baseline and 1, 2, 3, 4, 5 and 6 h after oral Phe loading (100 mg/kg body weight) (Hyland et al. 1997).

To evaluate specificity and sensitivity of these tests for the diagnosis of Segawa disease, the study was extended to 3 previously diagnosed patients (Table 1, cases 3–5) and 90 subjects with primary and secondary movement disorders not due to defects of biogenic amine metabolism who acted as control samples. In the whole, pterins in urine were assessed in 101 subjects, while 9 subjects with GCH1 mutations and 39 controls also underwent Phe loading test. No subject among patient and control population showed clinical and/or haematological features of concurrent infection disease when the urine sample was collected.

Statistical analysis. Quantitative data are presented as means $\pm$ standard deviations, and were analyzed by Mann–Whitney U test to assess differences between groups within each age class. Stepwise logistic regression analysis (backwards selection) was performed to select the variables associated with the diagnosis, separately in the two age groups. Potential diagnostic factors were biopterin, neopterin, Phe peak, Phe/Tyr at the second and third hour after Phe loading. The estimated Receiver Operating Characteristic (ROC) curve was then plotted for the selected model

Table 1 Genotype and biochemical phenotype in CGH1 mutated subjects enclosed in the study (see text for clinical details)

ID	Genotype	Sex	Age (years)	CSF HVA[a] (r.v.)[b]	CSF[a] 5-HIAA (r.v.)[b]	CSF NEO[a] (r.v.)[b]	CSF BIO[a] (r.v.)[b]	Urine NEO[a] (r.v.)[c]	Urine BIO[a] (r.v.)[c]	Ref.
1a	c.631-632 del AT [p.M211fs]	F	11	229 (148–434)	143 (68–115)	**7.7** (9.1–20.1)	11.8 (10.1–29.9)	**0.27** (0.30–1.84)	1.21 (0.53–2.05)	Pedigree 1: proband
1b	c.631-632 del AT [p.M211fs]	M	36					0.23 (0.19–0.91)	0.69 (0.28–0.86)	Pedigree 1: father of 1a
1c	c.671A>G [p.K224R]	M	8					**0.22** (0.30–1.84)	0.65 (0.53–2.05)	Pedigree 1: brother of 1a
1d	c.671A>G [p.K224R]	F	34					0.41 (0.19–0.91)	0.36 (0.28–0.86)	Pedigree 1: mother of 1a
2a	Exon 1 deletion	F	7	188 (137–582)	74 (68–220)	**5.4** (9.1–20.1)	11.1 (10.1–29.9)	**0.23** (0.30–1.84)	**0.42** (0.53–2.05)	Pedigree 2: proband
2b	Idem	F	36					**0.13** (0.19–0.91)	**0.16** (0.28–0.86)	Pedigree 2: mother of 2a
2c	Idem	M	5					**0.25** (0.30–1.84)	0.60 (0.53–2.05)	Pedigree 2: brother of 2a
2d	Idem	F	58					**0.13** (0.19–0.91)	**0.16** (0.28–0.86)	Pedigree 2: maternal grandmother 2a
3	c.68 C>T [p.P23L]	M	25	**81** (98–450)	45 (45–135)	**5.1** (9.1–20.1)	10.5 (10.1–29.9)	**0.11** (0.19–0.91)	**0.19** (0.28–0.86)	Leuzzi; unpublished case
4	c.671A>G [p.K224R]	M	17	**80** (98–450)	**42** (45–135)	**8.7** (9.1–20.1)	16.9 (10.1–29.9)	**0.16** (0.30–1.84)	**0.47** (0.53–2.05)	Leuzzi et al. 2002
5	c.262 C>T [p.R88W]	F	45					**0.14** (0.19–0.91)	0.4 (0.28–0.86)	Leuzzi; unpublished case

HVA homovanillic acid, *5-HIAA* 5-hydroxyndoleacetic acid, *NEO* neopterin, *BIO* biopterin, *BH4* tetrahydrobiopterin

[a] Pathological values are in bold

[b] Age-related reference values (r.v.) (nmol/l)

[c] Age-related reference values (r.v) (2.5–98 percentile of control values; mmol/mol creatinine)

and for the models including each of the selected variables, separately considered. The areas under the ROC curves were calculated as an overall measure of diagnostic efficiency of the selected criterion. Sensitivity (Se), specificity (Sp) and positive and negative predictive values (PPV and NPV, respectively) were computed. Finally, a curve representing the diagnostic test based on the selected model was determined.

All the statistical analyses were performed using STATA Statistical Software (Release 8.0).

Results

Table 1 shows *GCH1* genotype and biochemical phenotype of patients and asymptomatic carriers from the two families. Neopterin was low in CSF of both propositi (1a, 2a). The pattern of pterin excretion in urine (Table 1 cases 1a-2d) and Phe/Tyr ratio under oral Phe loading in symptomatic (4) and pre-symptomatic (4) carriers from the two families pointed both to GTP-GH deficiency.

The preliminarily exploration of the levels of neopterin and biopterin in urine from the controls showed they clustered differently before and over the age of 15 (Fig. 2a and b). Mean values of biopterin and neopterin in the "15–17 years" group were significantly different from those observed in the "0–14 years" group ($p = 0.0003$ and $p = 0.0137$, respectively), while they did not differ from those observed in the "18–74 years" group ($p = 0.8121$ and $p = 0.7979$, respectively). At the ratio Phe/Tyr under Phe loading test at second hour, no significant differences were observed in both comparisons ($p = 0.4857$ and $p = 0.1615$, respectively (Fig. 1c shows the distribution of Phe/Tyr at second hour). Therefore, for the following statistical analysis, GCH1 mutated subjects and controls were all grouped according to their age (pterin analysis: $59 < 15$ and $42 \geq 15$ years; Phe loading test: $23 < 15$, $25 \geq 15$ years).

Neopterin and biopterin concentrations (mmol/mol creatinine) in urine were, respectively, 0.66 ± 0.31 (range 0.22–1.84) and 0.99 ± 0.39 (range 0.47–2.54) in controls younger than 15, and 0.44 ± 0.20 (range 0.13–0.94) and 0.60 ± 0.16 (range 0.18-0.89) in those aged >15. In GCH1 mutated subjects, they were respectively 0.24 ± 0.02 (range 0.22–0.27) and 0.72 ± 0.34 (range 0.42–1.21), and 0.22 ± 0.11 (0.11-0.41) and 0.41 ± 0.19 (range 0.16-0.69) (GCH1 mutated subjects vs controls: <15 years neopterin $p = 0.0013$, biopterin $p = 0.1066$; ≥ 15 years neopterin $p = 0.0053$, biopterin $p = 0.0216$). Phe/Tyr ratio at seconnd and third hour after the loading of Phe were both significantly higher in patients than in controls (< 15 years: 5.59 ± 3.34 vs 1.91 ± 0.62, p = 0.0285; and 4.48 ± 3.88 vs 1.43 ± 0.47, $p = 0.0621$, respectively; ≥ 15 years:

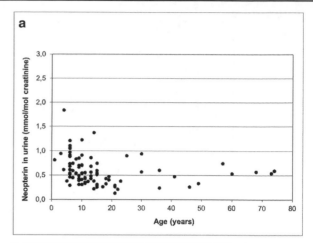

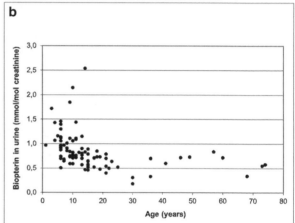

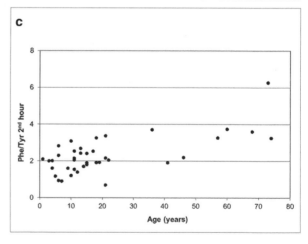

Fig. 1 Scatter plot of neopterin concentrations in urine vs age in controls ($n = 90$; $r = -0.19$; $p = 0.0748$) (**a**), biopterin concentrations in urine vs age in controls ($n = 90$; r $= -0.37$; $p = 0.0003$) (**b**) and Phe/Tyr ratio 2 h after Phe loading vs age in controls (**c**) ($n = 39$; $r = 0.68$; $p < 0.0001$)

7.27 ± 2.44 vs 2.73 ± 1.18, $p = 0.0006$; and 5.69 ± 2.36 vs 2.07 ± 0.84, p = 0.0007, respectively).

Backwards logistic regression identified Phe/Tyr at second hour and urine neopterin as significantly associated to *GCH1* gene alteration. The combined criterion appeared

Table 2 Results of the logistic regression analyses assessing the diagnostic value of Phe/Tyr ratio 2 h after Phe loading and neopterin concentration in urine

Area under the ROC curve	Phe/Tyr 2nd hour		Neopterin[o]		Phe/Tyr 2nd hour + Neopterin[o]	
	Age					
	<15 years (n = 23)	≥15 years (n = 25)	<15 years (n = 59)	≥15 years (n = 42)	<15 years (n = 23)	≥15 years (n = 25)
Se %	66.7	83.3	75.0	42.9	100.0	100.0
Sp %	100.0	94.7	98.2	94.3	100.0	100.0
PPV %	100.0	83.3	75.0	60.0	100.0	100.0
NPV %	95.2	94.7	98.2	89.2	100.0	100.0
Correctly classified %	95.7	92.0	96.6	85.7	100.0	100.0
Cut-off [a]	3.84	5.24	0.26	0.15	See Fig. 2a	See Fig. 2b
Logistic regression coefficients					See 'Results' and Fig. 2a	See 'Results' and Fig. 2b
Constant	−6.90	−7.59	11.48	1.48		
Independent variable	1.79	1.45	−44.12	−10.09		
p for the coefficient of the selected independent variable	0.145	0.022	0.053	0.022		

Se sensitivity, *Sp* specificity, *PPV* positive predictive values, *NPV* negative predictive values

[o] mmol/mol creatinine

[a] Cut-off values were obtained by logistic regression coefficients

to perform better in terms of specificity and sensitivity than the criteria based on Phe/Tyr at second hour and neopterin separately considered (Table 2). Considering both Phe/Tyr at second hour and neopterin values, the decision rule can be based on the following analytical expressions, obtained by the multivariate logistic regression (ROC analysis). The test is positive if:

"-27.47 - 464.38 * [neopterin] + 63.93 * Phe/Tyr 2 h > 0" for subjects younger than 15 years (Fig. 2a), and "-106.91 - 456.85 * [neopterin] + 53.48 * Phe/Tyr2h > 0" for subjects older than 15 years (Fig. 2b), where [neopterin] is the concentration of neopterin in urine (expressed in mmol/mol creatinine) and Phe/Tyr2h is the ratio between Phe and Tyr concentrations in dried blood spot (each expressed in μmol/l of blood) 2 h after Phe loading.

Discussion

The *GCH1* gene mutation is the most common cause of early onset DRD (Hagenah et al. 2005; Clot et al. 2009), involving about 85% of all patients with DRD (Furukawa 2004). More than 140 alterations have been so far reported on the GCH1 gene (www.hgmd.cf.ac.uk, accessed 17 November 2011). Two pathogenetic mutations occurred independently in family 1. p.M211fs was detected in the proband, who presented moderate mental retardation years before developing generalized choreoathetosis with diurnal

fluctuation. Both symptoms, and remarkably movement disorders, improved with L-Dopa/Carbidopa therapy. A mild dystonic phenotype was detected in her father. p.M211fs has been so far associated with a typical DRD presentation (Hagenah et al. 2005; Clot et al. 2009; Trender-Gerhard et al. 2009). Different to what happens in the recessive disorders of biogenic amine metabolism (Leuzzi et al. 2010), mental retardation is not part of DYT5a phenotype. It has been reported in the context of a complex presentation due to a 2.3 Mb deletion on chromosome 14q21-22 (family ITD612 in Clot et al. 2009) and in a single patient with a biochemically confirmed diagnosis but no alteration on the *GCH1* gene (Nagata et al. 2007). Two other members of Family 1 carried p.K224R mutation and were clinically normal in spite of some biochemical alterations suggesting GTP-CH deficiency (see below). p.K224R has been so far reported in three pedigrees associated with athetoid cerebral palsy (family Hu in Bandmann et al. 1998), myoclonic dystonia (Leuzzi et al. 2002) and adult-onset gait dystonia, rigidity and bradykinesia (Patient D97 in Garavaglia et al. 2004). Since the first report (Furukawa et al. 2000), few pedigrees harbouring gross deletion of the *GCH1* gene have been reported (Klein et al. 2002; Hagenah et al. 2005; Ohta et al. 2006; Clot et al. 2009). Their phenotype overlaps with that of patients with less severe alterations.

Since the availability of an aetiological treatment, the diagnosis of DYT5a should be considered for each

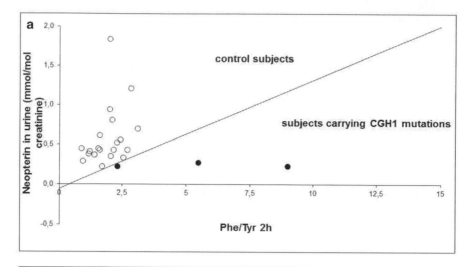

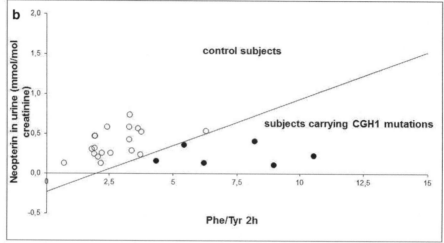

Fig. 2 Scatter plot of Phe/Tyr at second hour and neopterin values. The *white dots* represent controls (39) and the *black dots* patients (9) included in the ROC analysis. The curve divides the space into two areas, namely the area of control subjects above the curve and the area of subjects carrying GCH1 mutations below the curve. Each patient can be represented as a point in this space according to their values of neopterin and Phe/Tyr ratio 2 h after Phe loading: if the point lies in the area below the curve, the patient has a high probability of carrying a GCH1 mutations, whereas if they lie in the area above the curve, the

patient has a high probability of not carrying a GCH1 mutation. Decision rule can be expressed by the following analytical expressions. The test is positive if "-27.47 - 464.38 * [neopterin] + 63.93 * Phe/Tyr2h > 0" for subjects younger than 15 years (**a**) and "-106.91 - 456.85 * [neopterin] + 53.48 * Phe/Tyr2h > 0" for subjects older than 15 years (**b**), where [neopterin] is the concentration of neopterin in urine (expressed in mmol/mol creatinine) and Phe/Tyr2h is the ratio between Phe and Tyr concentrations in dried blood spot (each expressed in μmol/l of blood) 2 h after Phe loading

patient presenting with an early onset dystonia-parkinsonism syndromes. Tables 1 and 3 summarize the *GCH1* genotype and biochemical phenotype in the few cases in which both were examined. CSF examination was performed on 23 patients (including ours) from 16 pedigrees. Neopterin was low in 21/23 cases, biopterin or tetrahydrobiopterin in 18/22, HVA in 11/15, and 5-HIAA in 8/15. As a group, the subjects carrying GCH1 mutations showed also a lower level of neopterin in urine in comparison with the controls.

GTP-CH is the limiting enzyme for the synthesis of tetrahydrobiopterin, which is the cofactor of phenylalanine, tyrosine and tryptophan hydroxylases (Blau et al. 2001).

Based on this, a Phe loading procedure has been proposed (Hyland et al. 1997) and different values of Phe/Tyr 4 hours after the loading have been suggested as a critical threshold for the diagnosis: 4.5 (Saunders-Pullman et al. 2000), 7.5 (Bandmann et al. 2003), 5.25 (Saunders-Pullman et al. 2004). In spite of high sensitivity and specificity of the test, false negatives (Saunders-Pullman et al. 2004) and positives (Bandmann et al. 2003) were reported. Recent data suggest that the ratio obtained at the second hour may be more sensitive in young patients with DRD (Opladen et al. 2010; Lopez-Laso et al. 2006). The test reliability improves if the peak value of blood Phe exceeds 600 μmol/L and the plasma and blood spot biopterin variations during the loading are

Table 3 CGH1 genotype and biochemical phenotype in patients with Segawa disease from the literature

ID[a]	Genotype	Sex	CSF HVA[b] (r.v.)[c]	CSF 5-HIAA[b] (r.v.)[c]	CSF NEO[b] (r.v.)[c]	CSF BIO/BH4[b] (r.v.)[c]	Ref.
6	c.218C>A [p.A73D]	M	**145** (364–870)	**151** (155–359)	7 (5–53)	**<2** (20–61)	Opladen et al. 2010 (pt 1)
7	c.623 + 3insT [IVS5 + 3insT]	F	**95** (217–507)	**62** (75–203)	**3** (7–27)	**3** (20–49)	Ibidem (pt 2)
8	GCH1mutation	F	414 (220–560)	174 (90–237)	**4** (7–27)	**6** (20–49)	Ibidem (pt 6)
9	Heterozygous deletion of all 6 exons	M	**112** (217–507)	**34** (75–203)	**2** (7–27)	**2**(20–49)	Ibidem (pt 7)
10	c.547G>A [p.E183K]	F			**18.8** (25 ± 5.0)	**9.7** (25 ± 5.0)	Ikeda et al. 2009
11	c.64_65delGGinsAACC [p.G21fsX66]	F	**48** (115–455)	**20** (51–204)	**6** (10–31)	**n.d.** (18–53)	von Mering et al. 2008
12	c.159delG [p.W53fs]	F			**3.1** (22.6 ± 1.5)	**3.8** (20.6 ± 1.4)	Furuya et al. 2006
13	c.265C>T [p.Q89X]	M	**268** (334–906)	236 (170–420)	**2** (8–43)	**8** (11–39)	Lopez-Laso et al. 2006
14a	c.618del15bp [p.V206fs]	F			**8.5** (22.1 ± 7.0)	**8.2** (26.4 ± 8.5)	Ohta et al. 2006
14b	c.626 + 1G>A [IVS5 + 1G>A]	F			**3.9** (22.1 ± 7.0)	**1.3** (26.4 ± 8.5)	Ibid(pt 4)
14c	Exons 3-4 deletion	F			**8.6** (22.1 ± 7.0)	**15.7** (26.4 ± 8.5)	Ibid(pt 5)
15a	c.411del 37 bp [p.F138fs]	M	**142** (395 ± 56)	**50** (126 ± 22)	**1.1** (> 20)	**5.4** (> 20)	Hahn et al. 2001
15b	c.411del 37 bp [p.F138fs]	M	**85** (268 ± 24)	**43** (108 ± 15)	**4.8** (> 20)	**6.8** (>20)	Ibid(III:3)
15c	c.411del 37 bp [p.F138fs]	F	**447** (528 ± 75)	161 (133 ± 14)	**18.5** (> 20)	27.8 (> 20)	Ibidem(IV:6)
16a	c.671A>G [p.K224R]	F			**5.9** (7–65)		Bandmann et al. 1996; (Family Hu, II-3)
16b	c.671A>G [p.K224R]	F	91 (72–656)	65 (58–222)	**5.9** (7–65)	**6.6** (BH4) (9–40)	Robinson et al. 1999; (Family Hu, II-4)
16c	c.671A>G [p.K224R]	M	152 (72–256)	49 (58–222)	**4.3** (7–65)	16 (BH4) (9–40)	Ibid(Family Hu, II-5)
17a	c.344-1G>A [IVS1-1G>A]				**4.4** (13.0–38.3)	**7.1** (9.6–21.2)	Furukawa et al. 1996; (pt 2)
17b	c.341C>A [p.S114X]				**8.9** (13.0–38.3).	**5.9** (9.6–21.2)	Ibid(pt 3)

HVA homovanillic acid, *5-HIAA* 5-hydroxyndoleacetic acid, *NEO* neopterin; *BIO* biopterin, *BH4* tetrahydrobiopterin

nd: not detectable

[a] Numeration follows that of Table 1

[b] Pathological values are in bold

[c] r.v.: reference values (nmol/l)

monitored (Opladen et al. 2010, Saunders-Pullman et al. 2004).

We have showed that the analysis of neopterin in first morning urine leads to a correct classification in 96.6% (age < 15aa) and 85.7% (age > = 15 years) of 101 analyzed samples (Table 3) and that the evaluation of both Phe/Tyr ratio at the second hour and neopterin in urine leads to the highest specificity and sensitivity in order to detect young and adult subjects with GCH1 alterations, even though it cannot discriminate between symptomatic and pre-asymptomatic carriers (Hyland et al. 1999).

A pattern of Phe/Tyr ratio mimicking that found in GTP-CH deficiency may be found in the carriers of Phenylalanine Hydroxylase gene mutations (Driscoll and Hsai 1956; Guldberg et al. 1998), who, however, have a normal excretion of pterins in urine.

Caution should be adopted in interpreting urine and serum neopterin values in the presence of concomitant conditions potentially leading to T cell–mediated immune response activation (Huber et al. 1984) (such as viral infections, autoimmune diseases, malignancies, pregnancy, etc) (Werner-Felmayer et al. 2002): they could stimulate neopterin synthesis, so masking a possible partial defect of GTP-CH.

In conclusion, even though conducted in a restricted number of subjects, our study suggests a new simple, reliable and non-invasive approach to the diagnosis of DYT5a. It could be part of the diagnostic workup for patients presenting with idiopathic movement disorders of unknown origin.

Synopsis

The authors propose a rapid, simple and non-invasive method for the diagnosis of AD-DRD and suggest it as part of the diagnostic workup for patients presenting with idiopathic movement disorders of unknown origin.

Author Roles

Vincenzo Leuzzi: Conception, Organization, Execution of Research project; Writing of the first draft, Review and Critique of Statistical Analysis.

Claudia Carducci: Conception, Organization, Execution of Research project, Biochemical Studies; Review and Critique of Statistical Analysis.

Flavia Chiarotti: Review and Critique of experimental design, Statistical Analysis,

Daniela D'Agnano: Organization, Execution of Research project; Clinical data collection and patient follow-up; Review and Critique of Statistical Analysis.

Italo Antonozzi: Conception, Organization, Execution of Research project; Writing of the first draft, Review and Critique of Statistical Analysis

Maria Teresa Giannini: Organization, Execution of Research project; Clinical data collection and patient follow-up. Review and Critique of Statistical Analysis.

Carla Carducci: Organization, Execution of Research project; Molecular analysis; Review and Critique of Statistical Analysis.

Guarantor

Vincenzo Leuzzi

Competing Interests

The authors have declared that no competing interests exist.

Disclosure Information on Financial Support

No financial support.

Written Consent

A written consent of the legal substitute of the patients was obtained for genetic analysis.

References

Antonozzi I, Carducci C, Vestri L, Pontecorvi A, Moretti F (1988) Rapid and sensitive method for high-performance liquid chromatographic analysis of pterins in biological fluids. J Chromatogr 459:319–424

Bandmann O, Nygaard TG, Surtees R, Marsden CD, Wood NW, Harding AE (1996) Dopa responsive dystonia in British patients: new mutations of the GTP-cyclohydrolase I gene and evidence for genetic heterogeneity. Hum Mol Genet 5:403–406

Bandmann O, Valente EM, Holmans P et al (1998) Dopa-responsive dystonia: a clinical and molecular genetic study. Ann Neurol 44:649–656

Bandmann O, Goertz M, Zschocke J et al (2003) The phenylalanine loading test in the differential diagnosis of dystonia. Neurology 60:700–702

Blau N, Thony B, Cotton RGH, Hyland K (2001) Disorders of tetrahydrobiopterin and related biogenic amines. In: Scriver CR, Beaudet AL, Sly WS, Valle D, Childs B, Vogelstein B (eds) The metabolic and molecular bases of inherited disease. McGraw Hill, New York, pp 1725–1776

Chace DH, Millington DS, Terada N, Kahler SG, Roe CR, Hofman LF (1993) Rapid diagnosis of phenylketonuria by quantitative analysis for phenylalanine and tyrosine in neonatal blood spots by tandem mass spectrometry. Clin Chem 39:66–71

Clot F, Grabli D, Cazeneuve C et al (2009) Exhaustive analysis of BH4 and dopamine biosynthesis genes in patients with Dopa-responsive dystonia. Brain 132:1753–1763

Driscoll KW, Hsia DY (1956) Detection of the heterozygous carriers of phenylketonuria. Lancet 217(6957):1337–1338

Fink JK, Barton N, Cohen W, Lovenberg W, Burns RS, Hallett M (1988) Dystonia with marked diurnal variation associated with biopterin deficiency. Neurology 38:707–711

Furukawa Y, Nishi K, Kondo T, Mizuno Y, Narabayashi H (1993) CSF biopterin levels and clinical features of patients with juvenile parkinsonism. Adv Neurol 60:562–567

Furukawa Y, Shimadzu M, Rajput AH et al (1996) GTP-cyclohydrolase I gene mutations in hereditary progressive and Dopa-responsive dystonia. Ann Neurol 39:609–617

Furukawa Y, Shimadzu M, Hornykiewicz O, Kish S (1998) Molecular and biochemical aspects of hereditary progressive and Dopa-responsive dystonia. Adv Neurol 78:267–282

Furukawa Y, Guttman M, Sparagana SP et al (2000) Dopa-responsive dystonia due to a large deletion in the GTP cyclohydrolase I gene. Ann Neuro l47:517–520

Furukawa Y (2004) Update on dopa-responsive dystonia: locus heterogeneity and biochemical features. Adv Neurol 94:127–138

Furuya H, Murai H, Takasugi K et al (2006) A case of late-onset Segawa syndrome (autosomal dominant dopa-responsive dystonia) with a novel mutation of the GTP-cyclohydrase I (GCH1) gene. Clin Neurol Neurosurg 108:784–786

Garavaglia B, Invernizzi F, Carbone ML, et al (2004) GTP-cyclo-hydrolase I gene mutations in patients with autosomal dominant and recessive GTP-CH1 deficiency: identification and functional characterization of four novel mutations. J Inherit Metab Dis 27: 455–463

Guldberg P, Henriksen KF, Lou HC, Guttler F (1998) Aberrant phenylalanine metabolism in phenylketonuria heterozygotes. J Inherit Metab Dis 21:365–372

Hagenah J, Saunders-Pullman R, Hedrich K et al (2005) High mutation rate in dopa-responsive dystonia: detection with comprehensive GCHI screening. Neurology 64:908–911

Hahn H, Trant MR, Brownstein MJ, Harper RA, Milstien S, Butler IJ (2001) Neurologic and psychiatric manifestations in a family with a mutation in exon 2 of the guanosine triphosphate-cyclohydrolase gene. Arch Neurol 58:749–755

Hyland K, Fryburg JS, Wilson WG et al (1997) Oral phenylalanine loading in Dopa-responsive dystonia: a possible diagnostic test. Neurology 48:1290–1297

Huber C, Batchelor JR, Fuchs D et al (1984) Immune response-associated production of neopterin. Release from macrophages primarily under control of interferon-gamma. J Exp Med 160:310–316

Hyland K, Nygaard TG, Trugman JM, Swoboda KJ, Arnold LA, Sparagana SP (1999) Oral phenylalanine loading profiles in symptomatic and asymptomatic gene carriers with dopa-responsive dystonia due to dominantly inherited GTP cyclohydrolase deficiency. J Inherit Metab Dis 22:213–215

Ikeda T, Kanmura K, Kodama Y, Sawada K, Hunoi H, Hasegawa K (2009) Segawa disease with a novel heterozygous mutation in exon 5 of the GCH-1 gene (E183K). Brain Dev 31(2):173–175

Klein C, Hedrich K, Mohrmann K et al (2002) Exon deletions in the GCHI gene in two of four Turkish families with dopa-responsive dystonia. Neurology 59:1783–1786

Leuzzi V, Carducci CA, Carducci CL et al (2010) Phenotypic variability, neurological outcome and genetics background of 6-pyruvoyl-tetrahydropterin synthase deficiency. Clin Genet 77:249–257

Leuzzi V, Carducci C, Carducci C, Cardona F, Artiola C, Antonozzi I (2002) Autosomal dominant GTP-CH deficiency presenting as a dopa-responsive myoclonus-dystonia syndrome. Neurology 59:1241–1243

LeWitt PA, Miller LP, Levine RA et al (1986) Tetrahydrobiopterin in dystonia: identification of abnormal metabolism and therapeutic trials. Neurology 36:760–764

Lopez-Laso E, Ormazabal A, Camino R et al (2006) Oral phenylala-nine loading test for the diagnosis of dominant guanosine triphosphate cyclohydrolase I deficiency. Clin Biochem 39:893–897

Nagata E, Kosakai A, Tanaka K et al (2007) Dopa-responsive dystonia (Segawa disease)-like disease accompanied by mental retarda-tion: a case report. Mov Disord 22:1202–1203

Ohta E, Funayama M, Ichinose H et al (2006) Novel mutations in the guanosine triphosphate cyclohydrolase 1 gene associated with DYT5 dystonia. Arch Neurol 63:1605–1610

Opladen T, Okun JG, Burgard P, Blau N, Hoffmann GF (2010) Phenylalanine loading in pediatric patients with Dopa-responsive dystonia: revised test protocol and pediatric cut off values. J Inherit Metab Dis 101:48–54

Robinson R, McCarthy T, Bandmann O, Dobbie M, Surtees R, Wood NW (1999) GTP cyclohydrolase deficiency; intrafamilial varia-tion in clinical phenotype, including levodopa responsiveness. J Neurol Neurosurg Psychiatry 66:86–89

Saunders-Pullman R, Hyland K, Blau N et al (2000) Phenylalanine loading in the diagnosis of Dopa-responsive dystonia: the necessity for measuring biopterin. Ann Neurol 48:466

Saunders-Pullman R, Blau N, Hyland K et al (2004) Phenylalanine loading as a diagnostic test for DRD: interpreting the utility of the test. Mol Genet Metab 83:207–212

Segawa M, Nomura Y, Nishiyama N (2003) Autosomal dominant guanosine triphosphate cyclohydrolase I deficiency (Segawa disease). Ann Neurol 54(Suppl 6):S32–S45

Segawa M (2009) Autosomal dominant GTP cyclohydrolase I (AD GCH 1) deficiency (Segawa disease, dystonia 5; DYT 5). Chang Gung Med J 32(1):1–11

Schouten JP, McElgunn CJ, Waaijer R, Zwijnenburg D, Diepvens F, Pals G (2002) Relative quantification of 40 nucleic acid sequences by multiplex ligation-dependent probe amplification. Nucleic Acids Res 30:e 57

Trender-Gerhard I, Sweeney MG, Schwingenschuh P et al (2009) Autosomal-dominant GTPCH1-deficient DRD: clinical charac-teristics and long-term outcome of 34 patients. J Neurol Neuro-surg Psychiatry 80:839–845

Von Mering M, Gabriel H, Opladen T, Hoffmann GF, Storch A (2008) A novel mutation (c.64_65delGGinsAACC[p.G21fsX66]) in the GTP cyclohydrolase 1 gene that causes Segawa disease. J Neurol Neurosurg Psychiatry 79:229

Werner-Felmayer G, Golderer G, Werner ER (2002) Tetrahydrobiop-terin biosynthesis, utilization and pharmacological effects. Curr Drug Metab 3:159–173

Zirn B, Steinberger D, Troidl C et al (2008) Frequency of GCH1 deletions in Dopa-responsive dystonia. J Neurol Neurosurg Psychiatry 79:183–186

JIMD Reports
DOI 10.1007/8904_2012_145

β-Galactosidosis in Patient with Intermediate GM1 and MBD Phenotype

Tereza Moore · Jonathan A. Bernstein · Sylvie Casson-Parkin · Tina M. Cowan

Received: 10 January 2012 / Revised: 01 March 2012 / Accepted: 23 March 2012 / Published online: 22 April 2012
© SSIEM and Springer-Verlag Berlin Heidelberg 2012

Abstract A 5-year-old girl with clinical and biochemical phenotypes encompassing both GM1-gangliosidosis (GM1) and Morquio B disease (MBD) is described. Mild generalized skeletal dysplasia and keratan sulfaturia were consistent with a diagnosis of MBD, while developmental delay and GM1-specific oligosacchariduria were consistent with GM1 gangliosidosis. No observable β-galactosidase activity was detected in leukocytes, and two mutations, p.R201H (c.602G>A) and p.G311R (c.931G>A), were identified by gene sequencing. The R201H substitution has been previously reported in patients with both GM1 and MBD, and G311R is a novel mutation. Our patient represents a further example of the clinical heterogeneity that can result from mutations at the β-galactosidase locus.

Introduction

GM1-gangliosidosis (GM1) and Morquio B disease (MBD) are biochemically and phenotypically distinct lysosomal storage disorders caused by a deficiency in β-galactosidase (β-Gal, EC 3.2.1.23) activity, due to mutations in the *GLB1* gene (Suzuki et al. 2001). GM1 is a neurodegenerative disorder characterized by the accumulation of GM1 ganglioside in nervous tissue, and can be categorized into three phenotypic variants according to age of onset and severity of symptoms. The infantile form (Type I), shows early and rapid psychomotor deterioration, generalized central nervous system involvement, hepatosplenomegaly, cardiomyopathy, facial dysmorphism, and skeletal dysplasia (Suzuki et al. 2001). The late infantile/juvenile (Type II) and adult (Type III) forms display a progressive neurologic disease in childhood or early adulthood with localized skeletal and nervous system involvement, such as gait and speech disturbance (Suzuki et al. 2001). In contrast, patients with MBD retain neurological functions, but develop generalized skeletal dysplasia, keratan sulfaturia and corneal clouding (Suzuki et al. 2001). However, the clinical demarcation between GM1 and MBD can be obscured as in some patients displaying mental regression and the skeletal abnormalities of MBD (Giugliani et al. 1987; Mayer et al. 2009).

More than 130 sequence alterations in the *GLB1* gene have been identified so far, but our understanding of their effects on β-gal biosynthesis and function is still limited, and only a few may be predictive for one of the GM1 subtypes or MBD (Brunetti-Pierri and Scaglia 2008; Hofer et al. 2010). The tertiary structure of human GLB1 has recently been resolved, providing some insight into the bases of GM1 and MBD (Ohto et al. 2011). Based on crystallographic modeling, structural changes effecting β-gal protein folding, catalytic activity, substrate binding, and aggregation with lysosomal protective proteins have been predicted for various GLB1 mutations, but have not resulted in definitive phenotype classifications between the different diseases (Morita et al. 2009; Caciotti et al. 2011; Ohto et al. 2011).

We present a patient with intermediate clinical and biochemical phenotype between GM1 and MBD. The patient exhibits both skeletal disease and developmental delay, has deficient β-galactosidase enzyme activity, and has increased excretion of both GM1-specific oligosaccharides and keratan sulfate. Two mutations were identified by gene analysis, p.R201H (c.602G>A) and p.G311R (c.931G>A). R201H has been previously reported in both GM1 and MBD, and in the

Communicated by: Maurizio Scarpa

Competing interests: None declared

T. Moore (✉) · J.A. Bernstein · S. Casson-Parkin · T.M. Cowan
Department of Pathology, Stanford University, 3375 Hillview Avenue, Palo Alto, CA 94303, USA
e-mail: tereza@stanford.edu

intermediate GM1 and MBD phenotype (Caciotti et al. 2005; Paschke et al. 2001).G311R is a novel mutation involving a highly conserved amino acid residue, and is predicted to affect the β-gal catalytic site. This patient represents a further example of the problematic partition between GM1 and MBD, and highlights the need for further characterization of the *GLB1* gene and its substrate specificities.

Clinical Report

The patient is a 5-year-old girl with developmental delay who was referred to our metabolic clinic following the identification of cloudy corneas by ophthalmological exam. Her early cognitive development had been typical although she took her first steps at 18 months of age. At 2½ years of age her gait was noted to be slightly unsteady. At that time a prominence at the thoracolumbar junction was also observed. Generally, she experienced good health without hospitalization, major illness or surgery.

At the time of our evaluation the patient was of normal stature, weight and head circumference for age. No cardiac abnormalitites or organomegaly were appreciated on exam. Full range of motion was present in the fingers, wrists, elbows and knees. A developmental evaluation showed mild learning difficulties and evidence of processing delays. There was no recognized history of developmental regression. Neurological consultation noted mildly decreased muscle tone. An x-ray evaluation of the spine demonstrated mild beaking of L1 and L2. No radiologic abnormalities were observed in the long bones of the arm or in the wrist and hand.

Laboratory Analysis

Thin-layer chromatography (TLC) of urinary oligosaccharides showed an abnormal pattern characteristic of GM1. Interestingly, urine mucopolysaccharide TLC showed increased excretion of keratan sulfate, typical of patients with MBD, not GM1. β-gal activity in leukocytes using the artificial substrate 4 MU-β-D-galactopyranoside (Mayo Medical Laboratories) revealed no detectable enzyme activity, and molecular testing of the *GLB1* gene (Emory Genetics) identified two mutations: a previously reported p.R201H (c.602G>A) and a novel mutation p.G311R (c.931G>A). Molecular analysis of the patient's parents confirmed biparental inheritance; the father had a c.931G>A mutation and the mother a c.602G>A mutation.

Discussion

Deficiency in the activity of β-gal is expressed clinically and biochemically as GM1 and MBD. Classically, individuals with GM1 exhibit neurological deterioration and GM1-specific oligosacchariduria, without keratan sulfaturia, while those with MBD demonstrate normal intelligence, skeletal dysplasia, and keratan sulfaturia (Suzuki et al. 2001). β-gal catalyzes the removal of the β-linked galactose residue from its natural substrates, including ganglioside GM1, oligosaccharides, and keratan sulfate, and the differing clinical and biochemical phenotypes have been ascribed to the different substrate specificities of the mutant enzymes (Hofer et al. 2010).

Enzyme and mutation analysis have provided little insight into predicting the probable course and outcome of the disease. Residual β-gal enzyme activity of mutations expressed in vitro correlate fairly well with severity of disease, but the phenotypes of most compound heterozygous genotypes remain difficult to predict (Callahan 1999; Santamaria et al. 2007). Additionally, while it has been shown that most MBD patients carry a common mutation (p.W273L), only a few of the over 100 mutations known in GM1 can be related to a specific phenotype. Certain mutations have also been identified in both GM1 and MBD, and the same genetic assessment has been shown in patients who exhibited different symptoms, further complicating possible prognoses in these individuals (Kaye et al. 1997; Bagshaw et al. 2002; Roze et al. 2005; Santamaria et al. 2007; Caciotii et al. 2011).

The clinical and biochemical phenotypes of our patient obscure the lines between GM1 and MBD. Her mild skeletal abnormalities and keratan sulfaturia support the diagnosis of MBD, while her mild cognitive delay and GM1-specific oligosacchariduria support type II or III GM1. Whereas some mutations in *GLB1* can be predictive of the disease, the R201H mutation identified in this patient has been described in both GM1 type II/III and in MBD (Ishii et al. 1995; Kaye et al. 1997; Morrone et al. 2000; Paschke et al. 2001; Caciotti et al. 2005, 2011; Santamaria et al. 2006; Hofer et al. 2009). Interestingly, a patient homozygous for R201H was classified as MBD while other reports on heterozygous patients described more severe GM1 phenotypes (Santamaria et al. 2006; Morrone et al. 2000; Santamaria et al. 2007). Additionally, this mutation has been identified in two patients exhibiting skeletal changes of MBD together with neurological impairment (Caciotti et al. 2005; Paschke et al. 2001). Similar to our patient, both were heterozygous for the R201H mutation, and at least one was shown to have keratan sulfaturia and oligosacchariduria (Caciotti et al. 2005; Paschke et al. 2001). The R201H mutation is believed to cause a small conformational change, preventing the aggregation of β-gal with a lysosomal protective protein and resulting in its premature degradation (Morita et al. 2009; Ohto et al. 2011). In contrast to previous reports that describe residual enzyme activity for R201H, our patient had undetectable activity. The varied enzyme activity and phenotypes of individuals with R201H could be explained by the specific counter allele. The precursors derived from R201H would be prematurely degraded while those derived from the counter allele would predominantly reach the lysosomes and thus become determinant for the enzyme activity and phenotype.

The second mutation found in our patient (p.G311R) has not been reported in patients with MPS or GM1, nor has it been documented as a variant in the general population (www.ncbi.nih.gov/dbSNP; SIFT; PolyPhen). It is predicted to be causative of disease, however, due its location within the TIM barrel domain, a region of β-gal responsible for catalysis (Ohto et al. 2011). The Gly311 residue is also highly conserved between species and when replaced by a bulky, positive arginine residue in the vicinity of the catalytic site, may possibly weaken the enzyme-substrate interaction. The catalytic (Glu268, Glu188) and galactose binding (Tyr83, Asn187, Tyr333, Glu129, Ala128) residues of β-gal have been identified, but additional residues and/or domains may in addition be involved in the degradation of keratan sulfate (McCarter et al. 1997; Ohto et al. 2011). A mutation affecting the Tyr333 residue of the β-gal catalytic site (Y333H) has previously been observed in siblings with intermediate GM1 and MBD features together with keratan sulfaturia and GM1-specific oligosacchariduria (Giugliani et al. 1987; Mayer et al. 2009). Since our patient shows intermediate phenotype and both oligosaccharide and keratan storage, G311R could affect the ligand binding and/or catalytic activity of β-gal, and thus its specificity for both keratan sulfate and ganglioside GM1. Further characterization of mutant gene product, particularly for substrate specificity, will be required to clarify the pathogenesis as well as the prognoses of the diverse β-galactosidase disorders.

Synopsis

We present a case report of a 5-year-old girl with clinical and biochemical phenotypes encompassing both GM1-gangliosidosis (GM1) and Morquio B disease (MBD).

References

Bagshaw RD, Zhang S, Hinek A et al (2002) Novel mutations (Asn 484 Lys, Thr 500 Ala, Gly 438 Glu) in Morquio B disease. Biochim Biophys Acta 1588(3):247–253

Brunetti-Pierri N, Scaglia F (2008) GM1 gangliosidosis: review of clinical, molecular, and therapeutic aspects. Mol Genet Metab 94(4):391–396

Caciotti A, Donati MA, Bardelli T et al (2005) Primary and secondary elastin-binding protein defect leads to impaired elastogenesis in fibroblasts from GM1-gangliosidosis patients. Am J Pathol 167 (6):1689–1698

Caciotti A, Garman SC, Rivera-Colón Y et al (2011) GM1 gangliosidosis and Morquio B disease: an update on genetic alterations and clinical findings. Biochim Biophys Acta 1812 (7):782–790

Callahan JW (1999) Molecular basis of GM1 gangliosidosis and Morquio disease, type B. Structure-function studies of lysosomal beta-galactosidase and the non-lysosomal beta-galactosidase-like protein. Biochim Biophys Acta 1455(2–3):85–103

Giugliani R, Jackson M, Skinner SJ et al (1987) Progressive mental regression in siblings with Morquio disease type B (mucopoly-saccharidosis IV B). Clin Genet 32(5):313–325

Hofer D, Paul K, Fantur K et al (2009) GM1 gangliosidosis and Morquio B disease: expression analysis of missense mutations affecting the catalytic site of acid beta-galactosidase. Hum Mutat 30(8):1214–1221

Hofer D, Paul K, Fantur K, Beck M et al (2010) Phenotype determining alleles in GM1 gangliosidosis patients bearing novel GLB1 mutations. Clin Genet 78(3):236–246

Ishii N, Oohira T, Oshima A et al (1995) Clinical and molecular analysis of a Japanese boy with Morquio B disease. Clin Genet 48(2):103–108

Kaye EM, Shalish C, Livermore J, Taylor HA, Stevenson RE, Breakefield XO (1997) beta-Galactosidase gene mutations in patients with slowly progressive GM1 gangliosidosis. J Child Neurol 12(4):242–247

Mayer FQ, Pereira Fdos S, Fensom AH, Slade C, Matte U, Giugliani R (2009) New GLB1 mutation in siblings with Morquio type B disease presenting with mental regression. Mol Genet Metab 96 (3):148

McCarter JD, Burgoyne DL, Miao S, Zhang S, Callahan JW, Withers SG (1997) Identification of Glu-268 as the catalytic nucleophile of human lysosomal beta-galactosidase precursor by mass spectrometry. J Biol Chem 272(1):396–400

Morita M, Saito S, Ikeda K et al (2009) Structural bases of GM1 gangliosidosis and Morquio B disease. J Hum Genet 54 (9):510–515

Morrone A, Bardelli T, Donati MA et al (2000) Beta-galactosidase gene mutations affecting the lysosomal enzyme and the clastin-binding protein in GM1-gangliosidosis patients with cardiac involvement. Hum Mutat 15(4):354–366

Ohto U, Usui K, Ochi T, Yuki K, Satow Y, Shimizu T (2011) Crystal structure of human β-galactosidase: the structural basis of GM1 gangliosidosis and Morquio B diseases. J Biol Chem 2011 Nov 28 [Epub ahead of print]

Paschke E, Milos I, Kreimer-Erlacher H et al (2001) Mutation analyses in 17 patients with deficiency in acid beta-galactosidase: three novel point mutations and high correlation of mutation W273L with Morquio disease type B. Hum Genet 109 (2):159–166

Roze E, Paschke E, Lopez N et al (2005) Dystonia and parkinsonism in GM1 type 3 gangliosidosis. Mov Disord 20 (10):1366–1369

Santamaria R, ChabÄs A, Coll MJ, Miranda CS, Vilageliu L, Grinberg D (2006) Twenty-one novel mutations in the GLB1 gene identified in a large group of GM1-gangliosidosis and Morquio B patients: possible common origin for the prevalent p. R59H mutation among gypsies. Hum Mutat 27(10):1060

Santamaria R, ChabÄs A, Callahan JW, Grinberg D, Vilageliu L (2007) Expression and characterization of 14 GLB1 mutant alleles found in GM1-gangliosidosis and Morquio B patients. J Lipid Res 48(10):2275–2282

Suzuki Y, Oshima A, Namba E (2001) β-Galactosidase deficiency (β-galactosidosis) GM1 gangliosidosis and Morquio B disease. In: Scriver CR, Beaudet AL, Sly WS, Valle D (eds) The metabolic and molecular bases of inherited disease. McGraw-Hill, New York, pp 3775–3809

JIMD Reports
DOI 10.1007/8904_2012_146

RESEARCH REPORT

In Vivo Bone Architecture in Pompe Disease Using High-Resolution Peripheral Computed Tomography

**Aneal Khan · Zachary Weinstein · David A. Hanley ·
Robin Casey · Colleen McNeil · Barbara Ramage ·
Steven Boyd**

Received: 16 February 2012 /Revised: 02 April 2012 /Accepted: 5 April 2012 /Published online: 6 June 2012
© SSIEM and Springer-Verlag Berlin Heidelberg 2012

Abstract Pompe disease (lysosomal acid alpha-glucosidase deficiency) in adolescents and adults presents primarily with muscle weakness. Bone weakness is an under-recognized finding in patients with Pompe disease, but there is emerging evidence that loss of muscle function and mobility can lead to loss of mineral content and a higher risk of fracture. In addition to the mineral content, architecture is also important in determining the overall strength of the bone. We present the results of the longest longitudinal duration study to date using a novel application of high-resolution peripheral quantitative computed tomography (HR-pQCT) in four patients with Pompe disease over 4 years of observation during the normal course of their disease management. The subjects varied in treatment status with recombinant human alpha-glucosidase (rhGAA), use of anti-resorptive therapy (such as bisphosphonates), mobility and weight-bearing status, and the use of side-alternating vibration therapy. Our observations were that HR-pQCT can measure trends in mineral density and architecture over a long period of observation and may be an early indicator of the response to interventional therapies. In addition, a combination of decreased loading forces due to decreased mobility likely contributes to the compromise of bone integrity in Pompe disease. These trends can be reversed by applying increased loading forces such as vibration therapy and maintaining weight-bearing and mobility. We conclude that HR-pQCT can serve as a valuable tool to monitor bone health in patients with Pompe disease.

Communicated by: Ed Wraith

Competing interests: None declared

A. Khan (✉)
Department of Medical Genetics and Pediatrics, Alberta
Children's Hospital, University of Calgary, 2888 Shaganappi Trail
NW, Calgary, AB T3B 6A8, Canada
e-mail: khaa@ucalgary.ca

Z. Weinstein
University of Calgary, 2500 University Drive, NW, Calgary, AB
T2N 1 N4, Canada

D.A. Hanley
Departments of Medicine, Community Health Sciences and
Oncology, University of Calgary, Richmond Road Diagnostic
and Treatment Centre, 1820 Richmond Road SW, Calgary, AB
T2T 5 C7, Canada

R. Casey
Department of Medical Genetics and Pediatrics, Alberta
Children's Hospital, University of Calgary, 2888 Shaganappi Trail
NW, Calgary, AB T3B 6A8, Canada

C. McNeil
Alberta Children's Hospital, 2888 Shaganappi Trail NW, Calgary,
AB T3B 6A8, Canada

B. Ramage
Riddell Movement Assessment Centre, Alberta Children's
Hospital, Department of Pediatrics, University of Calgary, 2888
Shaganappi Trail NW, Calgary, AB T3B 6A8, Canada

S. Boyd
Schulich School of Engineering, University of Calgary,
2500 University Drive, NW, Calgary, AB T2N 1 N4, Canada

Introduction

Pompe disease (lysosomal acid alpha-glucosidase deficiency; OMIM 606800) is a rare autosomal recessive disorder that presents with muscle weakness with a natural history of progression to respiratory failure. The overall prevalence is 1:40,000 (Kishnani et al. 2006). The severe presentation, termed "the infantile" form, is accompanied by hypertrophic cardiomyopathy and is fatal in the majority of subjects in their first year of life without treatment (van den Hout et al. 2003). Pompe disease presenting after infancy, termed "the juvenile" or "late-onset" forms, presents at any age with neuromuscular weakness progressing to immobility and need for assisted ventilation (Hagemans et al. 2005). The standard treatment for Pompe disease involves biweekly infusions of recombinant human alpha-glucosidase (Kishnani et al. 2007).

There is emerging evidence that bone mass is compromised in subjects with Pompe disease and can lead to fractures (Papadimas et al. 2011a). There is no evidence currently to suggest alpha-glucosidase deficiency directly affects bone metabolism, but rather a combination of effects such as the degenerative myopathy from Pompe disease, loss of weight-bearing ability, and deconditioning with subsequent unloading of forces on the bone leading to secondary loss of bone mass (van den Berg et al. 2010). Although dual x-ray absorptiometry (DXA) is the standard clinical method of measurement of bone mass and assessment of fracture risk in the general population, the use of DXA to assess bone strength in Pompe disease presents a number of challenges. Subjects can have scoliosis and/or orthopedic hardware which can create artifacts during scanning or subjects may spend a significant portion of their time in a wheelchair because of the proximal myopathy, but still have use of their limbs resulting in regional differences in loading forces on the bone. DXA does not measure a true volumetric bone density, but rather an areal density in g/cm^2 calculated by dividing the total bone mineral content by the projected area of the X-ray image of the region of interest. This provides a reasonable estimate of fracture risk in the average-sized older adult, but underestimates true bone density in individuals with smaller bones (e.g., children, subjects with short stature) (Faulkner et al. 1995). Furthermore, bone architecture, and not just mineral density, contributes to the overall strength of bone and is not assessed by DXA. Combining the architectural analysis with densitometry can explain up to 90 % of the mechanical structure-function relationships in bone (Goulet et al. 1994). We present a novel application of a noninvasive, in vivo method of assessing bone microarchitecture using three-dimensional high-resolution peripheral quantitative computed tomography (HR-pQCT) in subjects with Pompe disease (Boutroy et al. 2005; Ruegsegger et al. 1996). Validation of HR-pQCT for densitometry and morphological measurements has been performed (MacNeil and Boyd 2007) and tests in premenopausal women, postmenopausal women, and age-matched controls have been performed (Macdonald et al. 2011a). However, to our knowledge, this technology has not been used to examine osteopenia in lysosomal storage diseases such as Pompe disease (Boutroy et al. 2005; Hasegawa et al. 2000; Cortet et al. 1999).

Methods

Subjects

HR-pQCT scans were performed on an annual basis for 4 years in all of our subjects with Pompe disease during their routine clinical care. During the period of this study, these were all of the adult subjects with Pompe disease monitored in the province of Alberta, Canada. A summary of subject characteristics is provided in Table 1.

Subject A is a 42-year-old woman diagnosed at 37 years of age presenting with muscle weakness. She does not have cardiomyopathy, and does not need assistive ventilation. She can walk with difficulty but without walking aids. She does not have a history of fractures. She received 20 mg/kg alglucosidase-alfa (Myozyme®) every other week initiated at the start of HR-pQCT measurements. Her lumbar spine (L1-L4) t-score was 0.3 and left total hip t-score 1.4 at the beginning of this study. Plasma 25-hydroxy vitamin D levels determined on two occasions were 64.5 and 73.9 nmol/L and the subject was prescribed a supplement of 1,000 IU per day of vitamin D.

Subject B is a 19-year-old male who presented with muscular weakness at 5 years of age and was diagnosed with Pompe disease at 15 1/2 years of age. He has not had any fractures. He had received 20 mg/kg alglucosidase-alfa every other week. He can stand but he cannot walk and uses a wheelchair for his mobility. He was started on side alternating vibration therapy as he lost mobility (Vibraflex®, Galileo®, Home Edition, Novotec Medical, Pforzheim, Germany). A steel frame was constructed that allows him to pull himself out of his wheelchair with support, hold his weight using his arms and legs while standing on the vibration platform for 4 minutes at 20 Hz per day three times a week. He maintained this activity during the course of HR-pQCT measurements described below. During the last 6 months of observation, he was treated with vitamin D in addition to his enzyme therapy. He had severe scoliosis, spinal rods and therefore his lumbar spine DXA scores were difficult to interpret. His left total hip z-score ranged from −5.5 to −4.1 and his total body z-score −4.7 to −2.7 during the course of this study. These changes were commensurate with an increase in height as part of his growth during adolescence. The subject's vitamin D was 31.2 to 98.1 nmol/L during the course of this study and he was prescribed a vitamin D dose of 2,000 IU per day.

Subject C is a 17-year-old male who presented with hypotonia and hypertrophic cardiomyopathy at 11 months of age. He had several long bone fragility fractures secondary to minor trauma (left femur, left humerus, and left radius) beginning while he was under 13 years of age. Treatment with alglucosidase-alfa 20 mg/kg i.v. every

Table 1 Descriptive information for the four Pompe disease subjects

Age (y)	Sex	Genotype	Mobility	Enzyme replacement therapy duration prior to first pQCT scan (months)
42	F	IVS 1-13T>G 1912G>T	Ambulatory without aids	1
19	M	c.525delT c.-32-13T>G	Weight-bearing but non-ambulatory	13
17	M	1927G>A 2040G>A	Non-weight-bearing and non-ambulatory	36
34	F	c.-32-13T>G 2481 + 102_2046 + 31del	Ambulatory without aids	None

2 weeks was initiated at 14 years of age and a few weeks later he was started on intravenous pamidronate. Pamidronate infusions were started at monthly intervals (due to overlap with his enzyme infusions) at 1 mg/kg/dose for three consecutive days per infusion cycle and were discontinued 1 year before the first HR-pQCT scan. Two years prior to his first HR-pQCT scan, his rhGAA dose was increased from 20 to 40 mg/kg every other week. The higher dose was due to his enrollment in the AGLU03306 study (An Exploratory, Open-Label Study of the Safety and Efficacy of High Dose or High Dosing Frequency Myozyme® (alglucosidase alfa) Treatment in Patients with Pompe Disease Who Do Not Have an Optimal Response to the Standard Dose Regimen). The family elected to continue at this does once the study was completed. He had severe scoliosis and his DXA spinal scores were difficult to interpret. His right total hip z-score was −4.8 to −4.2 over the course of this study. The subject's vitamin D level ranged from 46.4 to 99.1 nmol/L during the course of this study and the subject was prescribed a dose of 2,000 IU per day of vitamin D. The lower plasma levels of vitamin D reflect periods of decreased compliance at various times. Twenty-four months prior to the first HR-pQCT scan, the subject was started on intravenous pamidronate and continued until 20 months after the first HR-pQCT scan (18 years of age). He was initially started on an "Osteogenesis Protocol" of 0.5 mg/kg on day 1 and then 1 mg/kg on days 2 and 3 followed by 1 mg/kg every month for the remainder of the duration of infusions because of difficulties in traveling for both the enzyme and bisphosphonate infusions.

Subject D is a 34-year-old female who was diagnosed at 30 years of age. She has not had any fractures. She can ambulate but does have proximal muscle weakness, slow walking speed, and difficulty getting from a sitting to standing position without support. She was treated with 20 mg/kg alglucosidase-alfa every other week and alendronate sodium 70 mg plus 5,600 IU vitamin D_3 (Fosavance®) during the course of these observations. She started using a vibration platform (Vibraflex®, Galileo®, Home Edition, Novotec Medical, Pforzheim, Germany) after her first scan which added loaded forces to her lower limbs but not her upper limbs since she was able to balance on the platform without the use of her arms (Khan et al. 2009). Her lumbar spine (L1-L4) t-score ranged from 0.2 to 0.4 and her left total hip t-score ranged from −2.4 to −2.2 over the course of this study. The subject's vitamin D levels ranged from 78.2 to 83.9 nmol/L during the course of the study while taking an additional supplement of 1,000 IU per day (the Fosavance also contained vitamin D).

Measurements

HR-pQCT measurements. Scans were performed by HR-pQCT (XtremeCT, Scanco Medical, Brüttisellen, Switzerland) at two standard skeletal sites including the distal radius and the distal tibia (Boutroy et al. 2005). The subjects' arm or leg was supported in the scanner by a carbon cast to reduce subject motion. An initial x-ray ("scout view") was taken providing a basis to precisely manually identify the site for 3D scanning. The total time for each 3D scan was less than 3 min and resulted in 110 slices representing an axial section 9.02 mm long with a nominal isotropic resolution of 82 mm (60 kVp, 1000 μA, 100 ms integration time). The effective patient dose from a single scan is less than 3 μSv.

For each scan, a region of analysis (ROI) was defined at the periosteal surface for each slice using a semiautomated software routine, and subsequently the data was submitted for automated densitometry and morphometry evaluation (IPL v4.3, Scanco Medical). The main results from the analysis reported in equivalent mg/cm^3 of hydroxyapetite for the entire scanned region include total density in a scanned region (D100, mg/cm^3), cortical density (Dcort, mg/cm^3), and trabecular density (Dtrab, mg/cm^3). Additionally, morphometric parameters included the trabecular number (Tb. N, mm^{-1}) and cortical thickness (Ct.Th, mm). The trabecular number is determined based on a direct analysis of the 3D data and then trabecular thickness and separation

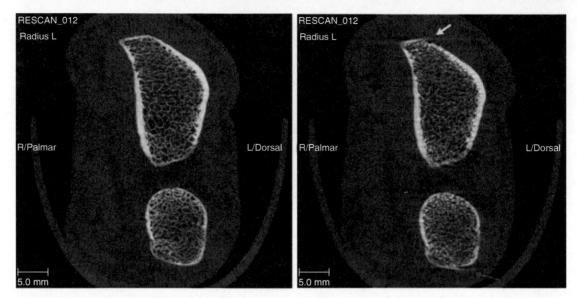

Fig. 1 Assessment of scan quality. The scan on the left shows no motion artifacts. The scan on the right shows radial whisping (*yellow arrow*) and whisping with cortical irregularities of the ulna (*right arrow*)

are derived (MacNeil and Boyd 2007). All subjects were scanned every 6 months during the actual course of their treatment.

Each scan was assessed according to manufacturer guidelines for quality on a scale of 0–4 for motion artifacts and verified by a second, independent assessment (0 = no motion artifact, 1 = slight whisping but no cortical discontinuity, 2 = whisping but no cortical discontinuities, 3 = high whisping and minor cortical discontinuities, and 4 = high whisping and major cortical discontinuities), and the entire scan discarded if any one of the images scored 3 or higher (Fig. 1). Individuals performing and reading the scans were blinded to the treatment details of the subjects.

Results

Figure 2 shows representative images of the left distal radius using HR-pQCT. Qualitative inspection of the scans shows how the underlying cortical thickness and trabecular number are reduced in subjects that are non-ambulatory, subjects B and C, compared to the subjects that were ambulatory, subjects A and D.

Analysis of the trends in mineral density and morphologic bone architecture are summarized in Fig. 3. For each subject, a 4-year interval is presented while they were monitored using HR-pQCT. The ordinate uses the same scale for each subject for the representative measurement. These trends were then analyzed with respect to time on standard therapy and differences in loading versus unloading of the limbs as follows:

(a) Changes in mineral density in unloaded limbs
    In subjects A, C, and D, the upper limbs were not exercised (hence no additional loading forces) and therefore these limbs show the natural course of bone density in patients being treated with rhGAA and vitamin D. Figure 3 shows the change in D100 over time at the radius. Subjects A and D, both ambulatory women on standard therapy, show continuous reductions in D100 over time. In comparison, subject C, treated with bisphosphonates prior to HR-pQCT scanning, showed an attenuated decline in D100. All three of these cases were treated with enzyme replacement therapy, but without any additional loading of the upper limbs. These three cases suggest that drug therapy alone was not sufficient to reverse the loss of bone mineral density in a non-weight-bearing limb.

(b) Changes in Mineral Density with Loading of the Limbs
    We analyzed compartmental changes in mineral density. Subject B had an external loading force applied using a vibration platform during which he supported his weight with both his hands and legs using a metal frame. Increases in D100 appear to be the result mostly of increases in cortical bone density (Dcomp) and, to a lesser degree, trabecular density (Dtrab), showing a 22.4 % improvement at the left radius in D100 and 19.5 % in Dcomp at the left tibia with comparable changes on the right side as well (26.3 % and 12.3 %, respectively).

(c) Changes in Mineral Density and Architecture
    Subject B shows that in addition to mineral density, cortical thickness and trabecular number both show increases when loading forces are added.

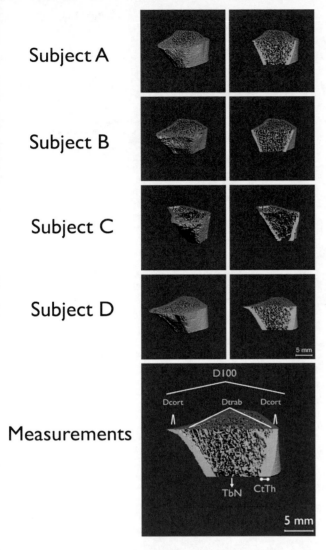

Subject A

Subject B

Subject C

Subject D

Measurements

Fig. 2 *Top.* Three-dimensional HR-pQCT imaging of left radius representing a 9.02 mm axial region located 9.5 mm from the distal end of the radius. Left column shows the entire scan and the right column shows the interior microarchitecture from the same scan data. *Bottom.* Measurements. Graphical representation of average density at the radius (D100), density of compact bone (DComp) and trabecular bone (Dtrab), and bone architecture represented by cortical thickness (CtTh) and trabecular number (TbN)

(d) Changes in Mineral Density in Loaded Versus Unloaded Limbs

The two ambulatory subjects are A and D. Both subjects were on alglucosidase-alfa and vitamin D but subject D was started on a vibration platform, standing only, without any extra force applied on her arms. Therefore, subject A did not use any additional loading force and subject D applied a loading force on the lower extremity but not on the upper extremity. Both subjects A and D had normal peripheral limb bone density and the use of vibration on the lower extremity in subject D showed no improvement when

compared to subject A. However, in subject B, who is weight-bearing but not ambulatory and had decreased peripheral limb bone density, there was a sustained improvement in mineral density and architecture using vibration in addition to enzyme replacement and vitamin D.

Discussion

These results are observations on a convenient sample of subjects under routine clinical care and treatment of their Pompe disease. There are no studies published using HR-pQCT to study bone architecture in subjects with Pompe disease and few longitudinal studies using HR-pQCT from osteopenia due to other causes (Burghardt et al. 2010; Rizzoli et al. 2010; Macdonald et al. 2011b). This study is the longest duration longitudinal study to date by HR-pQCT. This study also includes the first observations after using side alternating vibration therapy on bone architecture using HR-pQCT in two subjects (subjects B and D). The objective of the study was to pilot the use of HR-pQCT in the study of bone density and architecture in subjects with Pompe disease as an adjunct to other measures, such as plain roentgenography and DXA, in assessing their bone quality. HR-pQCT is an emerging technique available only in a small number of academic centers (currently four in Canada: Vancouver, Calgary, Saskatoon, and Toronto). Validation of the use of HR-pQCT on a larger set of subjects with Pompe disease would be valuable. Each center in Canada typically looks after only a few subjects with Pompe disease, therefore, collaboration with larger groups of subjects would provide a broader assessment of bone health in subjects with Pompe disease.

However, despite these limitations, even in this small set of subjects, we note some interesting observations after following them for 4 years: first, whether through a combination of reduced muscle mass, or decreased weight-bearing and mobility, the result of decreasing loading forces leads to continued declines in bone density and architecture in the peripheral limbs of subjects with Pompe disease. This trend did not appear to be reversed with a combination of enzyme replacement therapy, vitamin D and even oral bisphosphonates in some cases. Therefore, despite the benefits of ERT on respiratory function and mobility reported in clinical trials (Vielhaber et al. 2011; Papadimas et al. 2011b; van Capelle et al. 2010; Orlikowski et al. 2011), the treatment of subjects with Pompe disease should include load-bearing activities and physical activity in their management plan. Second, HR-pQCT can measure trends in mineral density and architecture that may be useful to indicate the direction of these changes before a

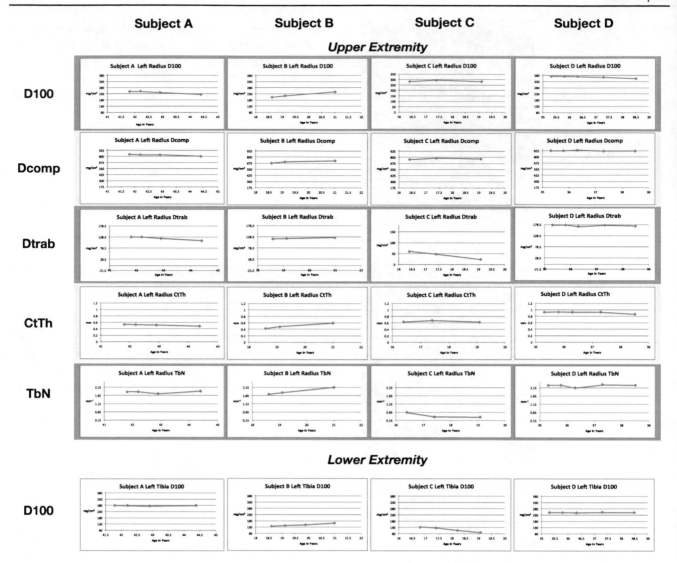

Fig. 3 Morphological trends in measurements of the left radius using HR-pQCT. Average density at the radius (D100), density of compact bone (DComp) and trabecular bone (Dtrab), and bone architecture represented by cortical thickness (CtTh) and trabecular number (TbN). Measurements were obtained for each limb distally but only data from the left radius is represented here due to space limitations

clinical endpoint, such as a fracture is noted. Third, the quality of the images on HR-pQCT should be assessed and trends should be evaluated over a long period of observation. We feel the overall message that needs to be stressed is to keep subjects with Pompe disease as mobile as possible in addition to their standard medical therapy. This is the first application of HR-pQCT to evaluate bone density and structure in subjects with Pompe disease and we feel there may be potential for clinical application of this technique in the future to monitor bone health.

The methods we used in obtaining HR-pQCT scans have been shown to have a precision of 0.7–1.5 % for total, trabecular, and cortical densities and 2.5–4.4 % for trabecular architecture (Boutroy et al. 2005; MacNeil and Boyd 2007). Overall, the changes we observed in our

subjects with Pompe disease over the course of 4 years are greater than the error range of HR-pQCT, and the changes in response to applying a loading force to the limbs exceeds that seen in shorter-term studies using alendronate in postmenopausal women (Burghardt et al. 2010). Nevertheless, there can be measurements in individual architectural parameters despite a good quality scan that should be interpreted with caution and verified by longitudinal follow-up. Hence, we feel the results from HR-pQCT should be evaluated using more than just two measurements.

This study was intended to be an observational study to pilot the use of HR-pQCT in subjects with Pompe disease. The treatments are varied since each subject was managed based on their individual clinical needs and therefore it is not intended to directly evaluate the effect of any particular treatment on bone architecture. All of our subjects had scoliosis and/or metallic prostheses that made the DXA scans difficult to interpret – in fact, this reason prompted us to use HR-pQCT to assess their bone. This study was not intended to directly compare DXA versus HR-pQCT in the assessment of bone health in patients with Pompe disease.

An HR-pQCT scan of the distal forearm gives an exposure of 3 μSv compared to a whole body DXA at 4.6 μSv and a background exposure of 7 μSv per day at sea level (Griffith and Genant 2008). Peak bone mass, the amount of bone at the end of skeletal maturation determined by DXA, is a major determinant of fracture risk in subjects who develop osteoporosis (Marshall et al. 1996). For each standard deviation reduction in the DXA T-score, there is an approximate doubling of the fracture risk (Cefalu 2004). However, bone mechanical strength is not dependent on mineral content alone. High-resolution peripheral quantitative computed tomography provides true volumetric density of both cancellous and cortical bone. In some subjects, cancellous bone is more responsive to interventions than cortical bone which appears to be the case in this study with trabecular indices showing more fluctuation than cortical measurements. Furthermore, in subjects with Pompe disease, angular deformities of the axial skeleton, disuse atrophy of the muscle and bone, and loss of ambulation and their effects on mineral density and architecture are not clear. We have demonstrated that HR-pQCT has potential as a noninvasive, in vivo method of evaluating changes in bone in subjects with Pompe disease.

Acknowledgments   Images in Fig. 1 are courtesy of Yves Pauchard, University of Calgary. We would like to thank Ion Robu for his engineering and technical expertise in the vibration training setup. This research was supported by funding from the Alberta Children's Hospital Research Foundation, Alberta Health Innovates Health Solutions (formerly Alberta Heritage Foundation for Medical Research), and through the support of Alberta Health Services.

Synopsis

High-resolution peripheral computed tomography can be used to monitor peripheral limb bone health in patients with Pompe disease.

References

Boutroy S, Bouxsein ML, Munoz F, Delmas PD (Dec 2005) In vivo assessment of trabecular bone microarchitecture by high-resolution peripheral quantitative computed tomography. J Clin Endocrinol Metab [Comparative Study Evaluation Studies Research Support, Non-U.S. Gov't]90(12):6508–6515

Burghardt AJ, Kazakia GJ, Sode M, de Papp AE, Link TM, Majumdar S (Dec 2010) A longitudinal HR-pQCT study of alendronate

treatment in postmenopausal women with low bone density: Relations among density, cortical and trabecular microarchitecture, biomechanics, and bone turnover. J Bone Miner Res [Randomized Controlled Trial Research Support, N.I.H., Extramural]25(12):2558–2571

Cefalu CA (March 2004) Is bone mineral density predictive of fracture risk reduction? Curr Med Res Opin [Review]20(3):341–349

Cortet B, Dubois P, Boutry N, Bourel P, Cotten A, Marchandise X (1999) Image analysis of the distal radius trabecular network using computed tomography. Osteoporos Int 9(5):410–419

Faulkner RA, McCulloch RG, Fyke SL, De Coteau WE, McKay HA, Bailey DA et al (1995) Comparison of areal and estimated volumetric bone mineral density values between older men and women. Osteoporos Int [Comparative Study Research Support, Non-U.S. Gov't]5(4):271–275

Goulet RW, Goldstein SA, Ciarelli MJ, Kuhn JL, Brown MB, Feldkamp LA (Apr 1994) The relationship between the structural and orthogonal compressive properties of trabecular bone. J Biomech [Research Support, U.S. Gov't, P.H.S.]27(4):375–389

Griffith JF, Genant HK (Oct 2008) Bone mass and architecture determination: state of the art. Best Pract Res Clin Endocrinol Metab [Review]22(5):737–764

Hagemans ML, Winkel LP, Van Doorn PA, Hop WJ, Loonen MC, Reuser AJ et al (March 2005) Clinical manifestation and natural course of late-onset Pompe's disease in 54 Dutch patients. Brain [Research Support, Non-U.S. Gov't]128(Pt 3):671–677

Hasegawa Y, Schneider P, Reiners C, Kushida K, Yamazaki K, Hasegawa K et al (2000) Estimation of the architectural properties of cortical bone using peripheral quantitative computed tomography. Osteoporos Int [Research Support, Non-U.S. Gov't] 11(1):36–42

Khan A, Ramage B, Robu I, Benard L (2009) Side-alternating vibration training improves muscle performance in a patient with late-onset pompe disease. Case Report Med 2009:741087

Kishnani PS, Hwu WL, Mandel H, Nicolino M, Yong F, Corzo D (May 2006) A retrospective, multinational, multicenter study on the natural history of infantile-onset Pompe disease. J Pediatr [Multicenter Study Research Support, Non-U.S. Gov't]148 (5):671–676

Kishnani PS, Corzo D, Nicolino M, Byrne B, Mandel H, Hwu WL et al (Jan 2007) Recombinant human acid [alpha]-glucosidase: major clinical benefits in infantile-onset Pompe disease. Neurology [Multicenter Study Randomized Controlled Trial]68(2):99–109

Macdonald HM, Nishiyama KK, Kang J, Hanley DA, Boyd SK (Jan 2011) Age-related patterns of trabecular and cortical bone loss differ between sexes and skeletal sites: a population-based HR-pQCT study. J Bone Miner Res [Research Support, Non-U.S. Gov't]26(1):50–62

Macdonald HM, Nishiyama KK, Hanley DA, Boyd SK (Jan 2011) Changes in trabecular and cortical bone microarchitecture at peripheral sites associated with 18 months of teriparatide therapy in postmenopausal women with osteoporosis. Osteoporos Int [Clinical Trial Research Support, Non-U.S. Gov't]22(1):357–362

MacNeil JA, Boyd SK (2007) Accuracy of high-resolution peripheral quantitative computed tomography for measurement of bone quality. Med Eng Phys 29(10):1096–1105

Marshall D, Johnell O, Wedel H (May 1996) Meta-analysis of how well measures of bone mineral density predict occurrence of osteoporotic fractures. BMJ [Meta-Analysis Research Support, Non-U.S. Gov't]312(7041):1254–1259

Orlikowski D, Pellegrini N, Prigent H, Laforet P, Carlier R, Carlier P et al (2011) Recombinant human acid alpha-glucosidase (rhGAA) in adult patients with severe respiratory failure due to Pompe disease. Neuromuscul Disord 21(7):477–482

Papadimas GK, Terzis G, Spengos K, Methenitis S, Papadopoulos C, Vassilopoulou S et al (Feb 2011) Bone mineral density in adult patients with Pompe disease. Bone [Comment Letter]48(2):417; author reply 8–9

Papadimas GK, Spengos K, Konstantinopoulou A, Vassilopoulou S, Vontzalidis A, Papadopoulos C et al (2011b) Adult Pompe disease: clinical manifestations and outcome of the first Greek patients receiving enzyme replacement therapy. Clin Neurol Neurosurg 113(4):303–307

Rizzoli R, Laroche M, Krieg MA, Frieling I, Thomas T, Delmas P et al (Aug 2010) Strontium ranelate and alendronate have differing effects on distal tibia bone microstructure in women with osteoporosis. Rheumatol Int [Randomized Controlled Trial Research Support, Non-U.S. Gov't]30(10):1341–1348

Ruegsegger P, Koller B, Muller R (Jan 1996) A microtomographic system for the nondestructive evaluation of bone architecture. Calcif Tissue Int [Research Support, Non-U.S. Gov't]58(1):24–29

van Capelle CI, van der Beek NA, Hagemans ML, Arts WF, Hop WC, Lee P et al (Dec 2010) Effect of enzyme therapy in juvenile patients with Pompe disease: a three-year open-label study. Neuromuscul Disord [Clinical Trial, Phase II Research Support, Non-U.S. Gov't]20(12):775–782

van den Berg LE, Zandbergen AA, van Capelle CI, de Vries JM, Hop WC, van den Hout JM et al (Sept 2010) Low bone mass in Pompe disease: muscular strength as a predictor of bone mineral density. Bone [Research Support, Non-U.S. Gov't]47 (3):643–649

van den Hout HM, Hop W, van Diggelen OP, Smeitink JA, Smit GP, Poll-The BT et al (Aug 2003) The natural course of infantile Pompe's disease: 20 original cases compared with 133 cases from the literature. Pediatrics [Research Support, Non-U.S. Gov't Review]112(2):332–340

Vielhaber S, Brejova A, Debska-Vielhaber G, Kaufmann J, Feistner H, Schoenfeld MA et al (2011) 24-months results in two adults with Pompe disease on enzyme replacement therapy. Clin Neurol Neurosurg 113(5):350–357

JIMD Reports
DOI 10.1007/8904_2012_153

CASE REPORT

A Case Study of Monozygotic Twins Apparently Homozygous for a Novel Variant of UDP-Galactose 4′-epimerase (GALE)

A Complex Case of Variant GALE

Ying Liu · Kristi Bentler · Bradford Coffee ·
Juliet S. Chhay · Kyriakie Sarafoglou ·
Judith L. Fridovich-Keil

Received: 13 February 2012 /Revised: 09 May 2012 /Accepted: 11 May 2012 /Published online: 1 July 2012
© SSIEM and Springer-Verlag Berlin Heidelberg 2012

Abstract Epimerase deficiency galactosemia is an autosomal recessive disorder that results from partial impairment of UDP-galactose 4′-epimerase (GALE), the third enzyme in the Leloir pathway of galactose metabolism. Clinical severity of epimerase deficiency ranges from potentially lethal to apparently benign, likely reflecting the extent of GALE enzyme impairment, among other factors. We report here a case study of monozygotic twins identified by newborn screening with elevated total galactose and normal galactose-1P uridylyltransferase (GALT). Follow-up testing revealed partial impairment of GALE in hemolysates but near-normal activity in lymphoblasts; molecular testing identified a missense substitution, R220W, apparently in the homozygous state. The twins were treated with dietary galactose restriction for the first 18 months of life. During this time, independent testing revealed concurrent diagnoses of Williams Syndrome in both twins, and cytomegalo-

virus (CMV) infection in one. Expression studies of R220W-hGALE in a null-background strain of *Saccharomyces cerevisiae* demonstrated a very limited impairment of V_{max} for UDP-galactose (UDP-Gal) and K_m for UDP-N-acetylgalactosamine (UDP-GalNAc), but a galactose challenge in vivo failed to uncover any evidence of impaired Leloir function. Similarly, both twins demonstrated normal hemolysate galactose-1-phosphate (Gal-1P) levels following normalization of their diets at 18 months of age. While these studies cannot rule out a negative consequence from some cryptic GALE impairment in a specific tissue or developmental stage, they suggest that the substitution, R220W, is mild to neutral, so that any GALE impairment in these twins is likely to be peripheral and therefore unlikely to be the cause of the negative outcomes observed.

Abbreviations

GALE	UDP-galactose 4′-epimerase
GALT	Galactose-1-phosphate uridylyltransferase
UDP-Gal	Uridine diphosphate galactose
UDP-GalNAc	Uridine diphosphate-N-acetyl-D-galactosamine
UDP-Glc	Uridine diphosphate gluctose
UDP-GlcNAc	Uridine diphosphate-N-acetyl-D-glucosamine

Communicated by: Alberto B Burlina

Competing interests: None declared

Y. Liu · J.S. Chhay · J.L. Fridovich-Keil (✉)
Department of Human Genetics, Emory University, School of Medicine, Room 325.2 Whitehead Building, 615 Michael Street, Atlanta, GA 30322, USA
e-mail: jfridov@emory.edu

K. Bentler · K. Sarafoglou
Division of Genetics and Metabolism, Department of Pediatrics, University of Minnesota Amplatz Children's Hospital, Minneapolis, MN 55455, USA

B. Coffee
Division of Medical Genetics, Department of Human Genetics, Emory University, School of Medicine, Atlanta, GA 30322, USA

Introduction

The galactosemias are a family of genetic disorders that result from impaired ability to metabolize galactose; these autosomal recessive conditions are caused by mutations that compromise either the expression or function of galactokinase (GALK, EC 2.7.1.6), galactose-1-phosphate uridylyltransferase (GALT, EC 2.7.7.12), or UDP-galactose 4′-epimerase (GALE, EC 5.1.3.2), the three enzymes of the Leloir pathway (Fridovich-Keil and Walter 2008). The symptoms and severity of these conditions vary in response to which enzyme is impaired and the extent of the impairment; other genetic and environmental factors remain poorly understood.

Epimerase deficiency (MIM# 230350), also called Type III galactosemia, results from partial impairment of GALE and is perhaps the least well understood of the galactosemias. Fewer than 10 patients with severe GALE impairment, resulting in a clinical presentation similar to that of classic transferase deficiency galactosemia, have been reported (Holton et al. 1981; Sarkar et al. 2010; Walter et al. 1999); these patients are said to have generalized epimerase deficiency (Fridovich-Keil et al. 2011). Similar to patients with classic galactosemia, patients with generalized epimerase deficiency respond well to dietary restriction of galactose, which prevents the potentially lethal acute symptoms, although long-term complications may persist (Henderson and Holton 1983; Sarkar et al. 2010; Walter et al. 1999).

Most patients with epimerase deficiency are asymptomatic in infancy and are identified through newborn screening in jurisdictions that measure both GALT activity and total galactose in every sample; the blood spots from these infants show elevated total galactose despite normal GALT (Fridovich-Keil et al. 2011). Follow-up enzymatic testing of hemolysates from patients with epimerase deficiency galactosemia demonstrates partial to profound loss of GALE activity. Enzymatic testing of fibroblasts or transformed lymphoblasts reveals a broad range of impairment from essentially normal GALE activity down to 15–20% of normal (Gitzelmann and Steimann 1973; Gitzelmann et al. 1976; Mitchell et al. 1975; Openo et al. 2006). Individuals with normal or near-normal GALE activity in cell types other than red blood cells are said to have peripheral epimerase deficiency, and are believed to remain asymptomatic (Fridovich-Keil et al. 2011; Gitzelmann et al. 1976). Individuals who show partial impairment of GALE activity in cell types other than red blood cells are said to have intermediate epimerase deficiency (Openo et al. 2006); the long-term outcomes for these patients remain unknown as nearly all are lost to follow-up at an early age (Alano et al. 1998; Fridovich-Keil et al. 2011; Quimby et al. 1997; Wohlers et al. 1999).

More than 20 different, ostensibly causal genetic variants have been identified in the GALE loci of patients with biochemically confirmed epimerase deficiency; a subset of these alleles and the GALE proteins they encode have been studied in vitro and/or in vivo, revealing a range of degrees and mechanisms of impairment (Alano et al. 1998; Bang et al. 2009; Chhay et al. 2008b; Henderson et al. 2001; Maceratesi et al. 1996, 1998; Openo et al. 2006; Park et al. 2005; Quimby et al. 1997; Thoden et al. 2001b; Timson 2005; Wohlers and Fridovich-Keil 2000; Wohlers et al. 1999). Some variant GALE proteins are catalytically impaired, while others demonstrate compromised stability, at least under defined laboratory conditions. While the relationship between GALE activity and galactose-sensitivity has been defined in yeast and *Drosophila* models (Sanders et al. 2010; Wasilenko and Fridovich-Keil 2006), the relationship between GALE activity level and outcome severity in patients remains a point of speculation.

We report a case study on the discordant clinical, biochemical, and genetic phenotypes of monozygotic twins with partial GALE deficiency identified by newborn screening on the basis of normal GALT and elevated total galactose levels. Both twins were apparently homozygous for a novel variant of GALE (R220W) and had multiple concurrent abnormalities including severe vitamin D–deficiency rickets, moderate bilateral sensorineural hearing loss, and genomic deletions consistent with Williams syndrome. Our report illustrates the complexity of comorbidities that can exist in a single patient and serves as a reminder that not every genetic variant is causal of the clinical abnormalities present.

Materials and Methods

Study subjects: The twins, designated here as FKE065 and FKE066, were ascertained by referral from their metabolic nurse practitioner; informed consent was obtained in accordance with Emory University Institutional Review Board Protocol 618–99. Control samples were ascertained as anonymous blood samples from nongalactosemic individuals, also in accordance with IRB protocol 618–99. The hemolysate biochemical data listed here were generated in clinical labs, as noted.

Mutational analysis of the hGALE locus: Mutational analysis of the *hGALE* loci in both twins was performed in a CLIA-approved clinical laboratory by direct sequencing of PCR-amplified fragments of genomic DNA representing all coding exons and their immediate flanking intronic sequences. Each amplicon was sequenced in both the forward and reverse directions and nucleotide changes were interpreted using the Human Genome Mutation Database (HGMD) and dbSNP database. The functional significance of the R220W substitution in hGALE was estimated using three software systems: SIFT (http://sift.jcvi.org/, Ng and Henikoff 2003), PANTHER (Thomas et al. 2003), and Polyphen 2 (http://genetics.bwh.harvard.edu/pph2/, Adzhubei et al. 2010).

Lymphoblast studies: EBV-transformed lymphoblasts were prepared from patient and control blood samples as described previously (Neitzel 1986). Transformed lymphoblasts were maintained in RPMI-1640 medium (Hyclone) containing glucose (2 g/L) and L-glutamine (0.3 g/L) and supplemented with penicillin (100 U/mL), streptomycin (100 mg/mL), 25 mmol/L Hepes, and 10% (v/v) fetal bovine serum (FBS) (Gibco/Invitrogen, Carlsbad, CA, USA). All cells were maintained at 37 °C in a humidified 5% CO_2 incubator (NuAire, Plymouth, MN, USA).

GALT and GALE enzyme analyses of lymphoblast cell lysates: Lymphoblast cell cultures were harvested and lysates prepared and analyzed for GALT and GALE activities as described previously (Openo et al. 2006). Enzyme activity was defined in units of pmol product produced/min/μg soluble protein.

Plasmids and yeast strains: The R220W substitution was re-created by site-directed mutagenesis of a wild-type human GALE coding sequence within the context of a centromeric yeast expression plasmid (MM33) that has been previously described (Chhay et al. 2008b). Mutagenesis was carried out using the Quick-change system (Stratagene, Inc.) according to the manufacturer's instructions using the following primer sequences (lower case letters indicate the mutation to be created): hGALE.R220W.f1 5′ GTGGCGATCGGGC-GAtGGGAGGCCCTGAATGTC 3′ and hGALE.R220W.r1 5′ GACATTCAGGGCCTCCCaTCGCCCGATCGCCAC 3′. The entire hGALE open reading frame of the resulting plasmid (pMM33.hGALE.R220W) was confirmed by dideoxy sequencing. The corresponding positive (wild-type hGALE) and negative (plasmid backbone only) plasmid controls have been reported previously (Chhay et al. 2008b; Wohlers et al. 1999).

All yeast strains used in this study were derived by transformation of JFy3835, a *GAL10*-null haploid strain of *Saccharomyces cerevisiae* that lacks endogenous GALE and has been described reviously (Chhay et al. 2008b). For in vitro biochemical assays, yeast cultures were maintained in standard synthetic medium containing 2% dextrose (SD) at 28 °C. For growth curves, yeast were cultured in standard synthetic medium containing 2% glycerol and 2% ethanol (SGE) at 30 °C, with or without the addition of galactose, as indicated.

Enzyme assays of yeast soluble lysates: Yeast proteins were extracted from cells harvested from SD cultures at OD_{600} between 0.8 and 1.2 by vigorous agitation with glass beads, as previously described (Chhay et al. 2008b). Soluble lysates were passed over P-6 Bio-Spin columns (Bio-Rad) to remove small metabolites prior to further analysis. The protein concentration in each sample was determined using the Bio-Rad protein assay reagent as recommended by the manufacturer using a standard curve of BSA. GALE and GALT enzyme activities were measured as described previously (Chhay et al. 2008b). Kinetic studies of each protein were performed in triplicate over a range of five different concentrations of substrate (UDP-Gal or UDP-GalNAc) with NAD^+ held constant, at 0.5 mM, and also in triplicate over a range of five different concentrations of NAD^+ with substrate held constant (UDP-Gal at 0.8 mM or UDP-GalNAc at 0.66 mM). Data were analyzed by SigmaPlot 11.0 software using the ligand-binding plot of the Michaelis–Menten equation ($1/V = K_m/(V_{max} [S]) + 1/V_{max}$).

Yeast growth assays in the presence of galactose: Yeast growth assays were performed as described previously (Chhay et al. 2008a).

Statistical analyses: Statistical analyses were performed using the JMP 8.0.1 software package. Data were analyzed using linear regression, one-way ANOVA, or *t*-test. The results were considered statistically significant if $P < 0.05$.

Case History

Infants FKE065 and FKE066 were delivered from a diamniotic dichorionic twin pregnancy in a 26-year-old, gravida 4 mother. The prenatal course was complicated by maternal iron deficiency anemia and growth restriction for fetus FKE065. Amniotic fluid testing for each fetus revealed a normal 46,XX female karyotype. Breech presentation led to cesarean section delivery of small-for-gestational-age twins with Apgar scores ≥6 at 37-5/7 weeks gestation. Polymerase chain reaction (PCR)-based DNA zygosity testing of the twins and their mother indicated a 98.2% probability of monozygosity. Parents are of indigenous Ecuadorian ancestry with no known consanguinity.

Neonatal complications were significant for failed hearing screenings for both twins, and heart murmur and mild jaundice for FKE066 that resolved without phototherapy. On the seventh day of life when abnormal newborn screening results for galactosemia were reported (Table 1), exclusively soy formula feedings were initiated. On the 11th day of life, during the initial metabolic consultation (Table 1), the twins had normal physical examinations and normal alanine aminotransferase levels (Table 2). Diagnostic testing revealed impairment of GALE enzymatic activity in hemolysates in the affected range (Table 1). At 9 weeks of age during metabolic follow-up, the twins were found to have new onset jaundice, cholestasis, conjugated hyperbilirubinemia, elevated liver transaminases without coagulopathy, gastroesophageal reflux, and failure to thrive, prompting prolonged hospitalizations (Table 2).

Throughout their hospitalizations, failure to thrive persisted despite high-caloric soy formula intake and fat-soluble vitamin supplementation. Neither twin had hepatomegaly. Biliary atresia was discounted by radionuclide imaging. Renal findings in both twins included generalized amino aciduria with echogenic kidneys and

Table 1 Biochemical, enzymatic, and molecular testing results

	FKE065	FKE066	Reference Ranges
GALT enzyme activity[a]	12.7	13.4	Control: ≥3.2 units/g Hb
Total galactose level[a]	38.3	30.3	Control: ≤18.0 mg%
Galactose-1-phosphate[b]	<0.1[c], 0.2[c], 0.1[c], 0.8[d], 0.1[d], <0.1[d]	0.2[c], 0.1[c], 0.1[c], 0.2[d], 0.4[d], 0.2[d], 0.5[d]	Control: 0.0–1.0 mg%
Galactitol, urine	22.6[c,e], 19.0[c,e], 32.7[c,e]	29.9[c,e], 26.0[c,e], 38.6[c,e]	Control: 0–94.7 mmol/mol Cr
GALE enzyme activity[b]	5.8,10.1	9.0, 7.9	Control: 17.1–40.1 μmol/h/g Hb
			Carrier: 12–20 μmol/h/g Hb
			Affected: 0–8 μmol/h/g Hb (Shin 1991)
GALT enzyme activity[b]	28.7	26.9	Control: 22.2–45.8 μmol/h/g Hb
Galactosemia isozymes	N/N	N/N	Control: N/N
GALK enzyme activity[b]	1.15	1.5	Control: 0.92–4.40 μmol/h/g Hb
GALE gene sequencing	c.658C>T (p.R220W)/c.658C>T (p.R220W)	c.658C>T (p.R220W)/c.658C>T (p.R220W)	Control: no mutations detected

[a] Blood spot newborn screen collected at 36 h of age

[b] Red blood cells

[c] Galactose-restricted diet (Gal-1P measured at <1, 8, and 13 months for FKE065 and at <1, 6, and 8 months for FKE066; galactitols collected twice at 2+ months and again at 3 months for each twin)

[d] Galactose-unrestricted diet (Gal-1P measured at 15, 21, and 34 months for FKE065 and 13, 15, 21, and 34 months for FKE066)

[e] Symptomatic cholestatic jaundice (9–16 weeks of age)

Table 2 Hepatic and endocrine laboratory results

	FKE065	FKE066	Normal reference range
Alanine aminotransferase	4[a,c], 44–215[b,c], 47[d]	8[a,c], 29–235[b,c], 15[d], 26[e]	0–50 u/L
Aspartate aminotransferase	79–435[b,c], 66[d]	62–399[b,c], 76[d], 33[e]	0–50 u/L
Alkaline phosphatase	683–1264[b,c],189[d]	822–2285[b,c], 190[d],164[e]	110–420 u/L
Bilirubin, conjugated	0.0–4.5[b,c], 0.0[d]	0.0–3.0[b,c], 0.0[d]	0–0.3 mg/dL
Bilirubin, total	1.3–7.6[b,c],<0.1[d]	0.7–5.9[b,c], 0.6[d], 0.4[e]	0.2–1.3 mg/dL
Bilirubin, delta	0.8–2.5[b,c], 0.0[d]	0.3–1.8[b,c], 0.5[d]	0.0–0.4 mg/dL
Gamma-glutamyltransferase	520[b,c]	Not measured	0–130 u/L
Serum calcium	7.2–10.0[b,c], 10.0[d]	7.7–9.6[b,c], 10.3[d]	9–11 mg/dL
Ionized calcium	3.2–5.7[b,c], 5.1[d]	3.6–5.8[b,c], 5.2[d]	5.1–6.3 mg/dL
Serum phosphorus	1.1–4.9[b,c], 5.0[d]	1.6–3.9[b,c], 5.9[d]	3.9–6.5 mg/dL
Parathyroid hormone, intact	151–641[b,c]	175–929[b,c]	12–72 pg/mL
25-hydroxyvitamin D, total	<10–< 36[b,c], <38[d]	<10–32[b,c],<34[d]	≥30 μg/L

[a] Asymptomatic, day 11 of age

[b] Acutely symptomatic 9–16 weeks of age

[c] Galactose-restricted diet (<1–4 months of age)

[d] Obtained at age 34 months (after 16 consecutive months on galactose-unrestricted diet without calcium or vitamin D supplementation)

[e] Obtained at age 56 months (after 38 consecutive months on a galactose-unrestricted diet)

interstitial nephritis. They also had severe bone demineralization and subacute fractures leading to diagnoses of severe vitamin D–deficient rickets (Table 2). The rickets diagnoses were attributed to probable severe maternal vitamin D deficiency, although maternal testing was unable to be completed. Osteogenesis imperfecta gene sequencing (COL1A1 and COL1A2) was later performed for FKE066 with negative results.

FKE066 had a cardiac murmur and mild supravalvar pulmonary stenosis. FKE065 also developed a cardiac murmur and was found to have supravalvar and branch pulmonary artery stenosis. Extensive infectious disease testing was negative for FKE066. FKE065 had high levels of cytomegalovirus (CMV) by DNA testing using PCR on multiple occasions, leading to liver biopsy at 12 weeks of age that suggested CMV hepatitis, and was treated with intravenous ganciclovir. Both infants had normal transcranial ultrasounds and normal ophthalmology examinations.

At hospital discharge, both infants showed healing rickets, correction of vitamin D deficiency after aggressive oral and IV therapy, resolution of cholestatic jaundice with ursodiol treatment and improvement of liver enzymes (Table 2).

In mid-infancy, both twins had high-resolution chromosome banding (46,XX), and array comparative genomic hybridization studies performed that demonstrated copy number losses consistent with deletions within band 7q11.23; FISH with a probe spanning the *LIMK* and *D7S613* loci confirmed the deletions and diagnoses of Williams syndrome. Subtle facial dysmorphism characteristic of Williams syndrome first became apparent for the infants at 8–11 months of age. By 16–21 months of age, both children had bitemporal narrowing, periorbital fullness, and full lips. FKE066 also had a coarse voice. By 5–8 months of age, the twins had hypotonia, brisk deep tendon reflexes, and language and gross motor delays. Neuropsychological evaluation at 10 months of age using the *Bayley Scales of Infant and Toddler Development* (Third Edition) documented pervasive developmental delays with cognitive, language, and motor age equivalencies in the 2–5 month range for both infants. Global developmental delays persisted; both twins sat unsupported at 13 months and walked independently at 28–29 months of age. At age 3 years, they remained largely nonverbal, hypotonic, and globally developmentally delayed.

GALE gene sequencing for each twin revealed apparent homozygosity for a previously unreported variant predicted by homology studies to be likely causative of GALE deficiency (Table 1). Their mother was later determined to be heterozygous for this sequence variant. Paternal genetic testing was unable to be performed. Galactose restriction was maintained throughout infancy as the GALE deficiency evaluation remained in progress, with galactose challenges undertaken at 13–15 months of age (Table 1). By 18 months of age, both children were on normal diets, unrestricted in galactose. Post-galactose challenge evaluations revealed normal eye examinations and no elevation of hemolysate Gal-1P for either twin (Table 1), and a normal liver ultrasound for FKE065.

At the time of this report, FKE065 has required 10 additional hospitalizations with subsequent bacteremia, gastrostomy tube placement (later removed), systemic hypertension requiring antihypertensive therapy (resolved), acute renal insufficiency (resolved), and marked thoracic scoliosis. FKE066 has required two additional hospitalizations for respiratory infections. Both children have moderate bilateral sensorineural hearing loss. They have been discharged from further metabolic follow-up.

Results

hGALE and hGALT Activity in Patient Lymphoblast Cells

To characterize the ostensible GALE impairment in these twins, we prepared EBV-transformed lymphoblasts from each and tested the resulting cell lysates for GALT and GALE activities (Fridovich-Keil et al. 2011; Openo et al. 2006). Three control lines (FKT901, FKT934, and GM) and one from a patient previously diagnosed with intermediate epimerase deficiency galactosemia (FKE084) were analyzed in parallel.

Standard GALE assays of lysates from FKE065 and FKE066 demonstrated no significant impairment relative to the three unaffected controls (Fig. 1a, $P = 0.4671$) while the affected lysate, FKE084, demonstrated only about 21% of control activity ($P < 0.0001$). Of note, GALE activities detected in lymphoblasts from the twins (FKE065 and FKE066) differed by close to a factor of 2; this range was also observed in lymphoblasts from different unaffected controls (Fig. 1a; Openo et al. 2006). As expected, the GALT activity levels in all six cell lines were indistinguishable ($P = 0.1986$, Fig. 1b).

Characterization of Human GALE-R220W Expressed in Yeast

To assess the functional significance of the R220W substitution in a more controlled setting, we recreated the variation by site-directed mutagenesis and expressed the resultant allele from a low-copy (CEN) plasmid in a null-background strain of *S. cerevisiae*, JFy3835 (Chhay et al. 2008b). Parallel cultures of yeast expressing wild-type human GALE (WT) or no GALE (bb only) served as positive and negative controls, respectively (Wohlers et al. 1999; Quimby et al. 1997).

Under normal assay conditions (Chhay et al. 2008b) using either UDP-Gal or UDP-GalNAc as substrate, hGALE-R220W demonstrated >50% wild-type activity (Fig. 2a and 2b). However, the apparent V_{max} values for the wild-type and hGALE-R220W proteins differed by about two-fold when the substrate was UDP-Gal (Table 3), and the apparent K_m values differed by about a factor of 2 when the substrate was UDP-GalNAc (Table 4). These

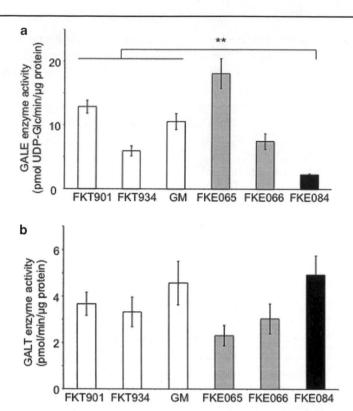

Fig. 1 *GALE and GALT enzyme activity levels in control and patient lymphoblast cells.* EBV-transformed lymphoblast cells cultured in complete RPMI-1640 medium with 10% FBS were collected by centrifugation and soluble lysates were prepared. FKT901, FKT934, and GM were derived from non-galactosemic controls (*white bars*). FKE065 and FKE066 represent the twins reported here (*gray bars*). FKE084 represents a patient with previously diagnosed intermediate epimerase deficiency (*black bar*). (**a**) GALE and (**b**) GALT enzyme activities were measured as described in Materials and Methods. Activity is expressed as mean $\pm$ SEM ($n \geq 4$) pmol/min/µg soluble protein

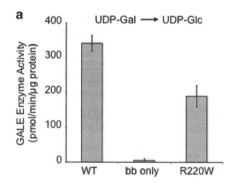

Fig. 2 *Enzyme activity of R220W-hGALE expressed in null-background yeast.* Colonies were inoculated from *GAL10-null* yeast transformed with plasmids encoding either wild-type hGALE (WT), no hGALE (backbone (bb) only), or R220W-hGALE (R220W). Cultures were expanded and harvested, and lysates prepared as described in Materials and Methods. GALE enzyme activity was monitored using either (**a**) UDP-Gal or (**b**) UDP-GalNAc as substrate. Activity is expressed as mean $\pm$ SEM ($n \geq 4$) pmol/min/µg soluble protein

V_{max} and K_m differences might explain why the apparent activity of the hGALE-R220W-substituted protein was reduced by close to twofold relative to the wild-type protein when measured using the standard assay (Fig. 2).

We also tested for possible impact of the R220W substitution on cofactor dependence of the enzyme activity (Frey 1996; Thoden et al. 1996, 2001a) by monitoring reactions performed under initial conditions of fixed substrate concentration and varying amounts of exogenous NAD$^+$. Although some patient mutations have been shown to disrupt hGALE interactions with NAD$^+$ (Quimby et al. 1997), we saw no evidence of such an impact by R220W (Fig. 3).

Table 3 Apparent K_m and V_{max} values of WT and R220W-hGALE proteins (UDP-Gal)

hGALE Derivative	Apparent UDP-Gal K_m (mean $\pm$ SEM mM)	Apparent UDP-Gal V_{max} (mean $\pm$ SEM pmol/ min/µg protein)
WT	0.314 $\pm$ 0.014	456.69 $\pm$ 6.35
R220W	0.274 $\pm$ 0.023	212.58 $\pm$ 5.49*

Assays to determine the K_m and V_{max} of UDP-Gal were performed at the concentration of 0.5 mM NAD$^+$. Kinetic constants were determined by fitting the data to a ligand-binding plot by SigmaPlot 11.0. All values are averages $\pm$ SEM ($n = 3$) of three independent analyses

*Indicates $P \leq 0.05$

Table 4 Apparent K_m and V_{max} values of WT and R220W-hGALE (UDP-GalNAc)

hGALE derivative	Apparent UDP-GalNAc K_m (mean $\pm$ SEM mM)	Apparent UDP-GalNAc V_{max} (mean $\pm$ SEM pmol/ min/µg protein)
WT	0.165 $\pm$ 0.038	114.39 $\pm$ 7.24
R220W	0.339 $\pm$ 0.045*	111.62 $\pm$ 5.47

Assays to determining the K_m and V_{max} of UDP-GalNAc were performed at the concentration of 0.5 mM NAD$^+$. Kinetic constants were determined by fitting the data to a ligand-binding plot by SigmaPlot 11.0. All values are averages $\pm$ SEM ($n = 3$) of three independent analyses

*Indicates $P \leq 0.05$

Galactose Sensitivity of Yeast Expressing hGALE-R220W

Finally, we measured the galactose-sensitivity of yeast expressing hGALE-R220W as their only GALE enzyme. Previously, we have demonstrated that the growth rate of yeast with compromised GALE activity is diminished in medium containing glycerol-ethanol as the carbon source when trace levels of galactose are spiked into the medium, and this effect is dose dependent (Ross et al. 2004; Wasilenko and Fridovich-Keil 2006). Here we tested the growth rates of yeast expressing wild-type hGALE (WT), no hGALE (bb only), or hGALE-R220W (R220W), each cultured in synthetic glycerol-ethanol medium without galactose vs. with 0.002% or 0.02% galactose added at t = 0 (Fig. 4). In medium lacking galactose, all three of the strains grew well (Fig. 4a), while in medium spiked with galactose, yeast expressing either wild-type hGALE or hGALE-R220W grew well, but yeast missing GALE did not (Fig. 4b, c). These data confirm that *hGALE-R220W* encodes a GALE enzyme that is predominantly, if not fully, functional in living yeast cells.

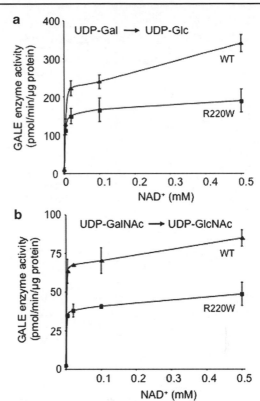

Fig. 3 *Effect of NAD$^+$ on GALE activity detected in lysates of yeast expressing wild-type vs. R220W-hGALE.* GALE assays were performed as described in Materials and Methods using soluble lysates of yeast expressing either wild-type hGALE or R220W-hGALE. (**a**) The initial concentration of substrate UDP-Gal was held constant at 0.8 mM; (**b**) the initial concentration of substrate UDP-GalNAc was held constant at 0.66 mM; NAD$^+$ concentration was varied as indicated. Activity is expressed as mean $\pm$ SEM ($n \geq 4$) pmol/min/ µg soluble protein

Discussion

The purpose of this study was threefold. First, we wanted to illustrate the complex challenges of elucidating the clinical phenotype of epimerase deficiency in patients with concurrent comorbidities. In this case, outcome was confounded by the presence of additional genetic (Williams syndrome) and environmental (CMV and vitamin D deficiency) issues; had there not been a pair of monozygotic twins, both of whom shared the same GALE genotype, but who demonstrated discordant outcomes for some symptoms, it might have been tempting to attribute more of the clinical complications to compromised GALE function.

Second, we wanted to address the question: How does one distinguish a functionally neutral or near-neutral variant from a functionally significant GALE mutation, especially when the genetic background of the patient may be distinct from that of publicly available control populations? That the variant found in this family had not been reported in control populations (e.g., HapMap) of predominantly European, Asian, or African

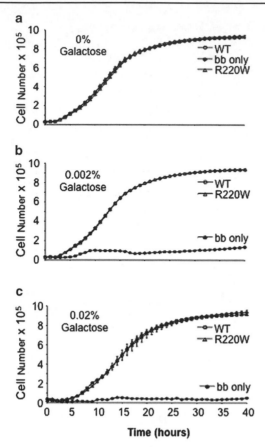

Fig. 4 *Yeast expressing R220W-hGALE demonstrate no apparent growth-sensitivity to galactose.* Cultures of null-background yeast expressing either wild-type hGALE (WT), no hGALE (bb only), or R220W-hGALE (R220W) were inoculated into the wells of 96-well plates at a density of 5×10^5 cells/mL at $t = 0$ in SGE-Ura synthetic medium supplemented with the indicated amount of galactose. Growth of each culture was monitored as described in Materials and Methods. Values plotted represent means $\pm$ SEM ($n = 9$). (**a**) All cultures grew well in the absence of galactose. (**b**) Addition of 0.002% galactose halted the growth of yeast lacking hGALE but not yeast expressing WT or R220W-hGALE. (**c**) Addition of 0.02% galactose halted the growth of yeast lacking hGALE but not yeast expressing WT or R220W-hGALE

ancestry may not be particularly informative given that the family traces their roots to Ecuador. The situation described here was further complicated by the observation that the twins are apparently homozygous for this novel variant. Testing of the father was never performed and it remains possible that he is a carrier of a *GALE* allele, such as a deletion, that escapes detection by our assay and therefore results in apparent homozygosity of the other allele. If the twins are truly homozygous for the R220W variant allele, it might be common in the relevant ancestral population. Alternatively, there might be some unrecognized shared ancestry between the parents, leading to concern that the twins might be homozygous at other loci, as well, and these cryptic homozygosities might underlie at least some of the apparent clinical abnormalities. Of note, consanguinity was a

known confounding factor in both affected families of the first patients reported with generalized epimerase deficiency galactosemia (Henderson and Holton 1983; Holton et al. 1981; Sardharwalla et al. 1988; Walter et al. 1999) who demonstrated a variety of developmental or other disabilities. A patient reported more recently (Sarkar et al. 2010), who also presented with acutely symptomatic epimerase deficiency in infancy, responded well to dietary restriction of galactose and apparently has not exhibited developmental delays; it is unknown whether there is consanguinity in that family.

Finally, we wanted to address the balanced relationship between in silico predictions of functional significance of amino acid substitutions, predictions from in vitro studies of recombinant proteins, and indicators of function in vivo in a yeast model system. As discussed below, in silico, in vitro, and in vivo studies can give disparate results, leading to the need for a judgment call that balances the weight of the accumulated clinical and laboratory data.

Arguments pointing toward clinically significant epimerase deficiency: The initial newborn screening result of the twins showed elevated total galactose at about twice the normal cut-off (Table 2). Repeat follow-up hemolysate GALE enzyme assays showed reduced activity within the affected range (Table 1) and apparent homozygosity for a novel variant of GALE (R220W) that in silico prediction tools classified as likely to be of functional significance. Polyphen-2 predicted this amino acid replacement as "probably damaging" with a score 1.00. The arginine at position 220 in the GALE protein is conserved among numerous mammalian species. In addition, arginine is conserved at position 220 in the amphibian *Xenopus* as well as in various species of fish indicating that this amino acid is conserved through many different vertebrates. Indeed, only chicken had a different amino acid, glutamine, at this position.

Arguments pointing away from functionally significant epimerase deficiency galactosemia: The neonatal presentation of the twins was unremarkable despite the fact that both infants were on a milk diet, that the lymphoblast GALE activity assay showed essentially normal values, and that the yeast expression studies demonstrated only very marginally lowered GALE activity in vitro and no apparent GALE defect in vivo. It is important to note that acute jaundice and other complications did not present until months after the diet had been switched to exclusively soy formula (Case History). The marginal elevation of total galactose in the newborn screens, coupled with the intermediate and variable hemolysate GALE results reported for both twins from follow-up studies, leaves open the question of the functional significance of these data. Indeed, that the twins were delivered at less than 38 weeks gestation and small for gestational age suggests that "immature liver" might have contributed to the newborn screening results (Fridovich-Keil et al. 2011; Ono et al. 1999). Finally, when taken off the galactose-restricted diet at 18 months, Gal-1P did not elevate in either twin, suggesting that the Leloir pathway in these infants was functioning, at least at that point.

Neutral variant or significant mutation? Novel missense variants at any genomic locus in a symptomatic patient are typically assessed for functional significance initially by two approaches: (1) querying prevalence of the variant in relevant affected and control populations and (2) in silico prediction programs that utilize multiple sequence alignments and structural information.

The twins reported here are apparently homozygous for an allele of GALE that has not been described previously; this suggests only that the allele may be uncommon in European and other studied populations. We have no information on the frequency of this allele among the indigenous peoples of Ecuador.

Due to the complex clinical picture of the twins (Williams syndrome and sensorineural hearing loss), additional steps were taken to test the functional significance of the identified missense variation; these included biochemical studies of transformed patient lymphoblasts, and in vitro and in vivo measures of GALE function in a null-background yeast strain engineered to express the relevant patient allele (R220W-hGALE). All of these measures demonstrated near-normal GALE activity, suggesting the R220W substitution does not significantly disrupt hGALE function, at least in non-peripheral cells. These results are consistent with a diagnosis of peripheral epimerase deficiency in FKE065 and FKE066. Some small differences (e.g., K_m or V_{max}) were detected in some assays between R220W-hGALE and the unaffected control; however, these differences were generally no more than a factor of 2, which is unlikely to impact clinical outcome in a recessive condition. Of course, we cannot rule out that the R220W substitution might have some significant but cryptic impact not tested for here, for example, restricted to a specific tissue, or time in development.

Acknowledgments We would like to thank the patients and their family for their willingness to share their story by participating in this research study. We also thank our colleagues at the University of Minnesota and at Emory University for many helpful conversations. This work was supported in part by funding from the National Institutes of Health (R01 DK059904 to JLFK).

One Sentence Summary

Detailed biochemical studies of a rare variant of human GALE, R220W, suggest that it is unlikely to account for the negative outcomes observed in the twins who carry it.

Contributions of the Individual Authors

Y Liu performed the majority of experiments presented, assembled the figures and tables, and wrote the first draft of the manuscript. K Bentler identified this family, assembled all of the clinical and some of the laboratory data, and wrote the manuscript section dealing with clinical presentation. B Coffee oversaw and interpreted the GALE genotyping and wrote the section of the manuscript relevant to those data. JS Chhay generated the hGALE-R220W yeast expression plasmid and also prepared and initially characterized the EBV-transformed lymphoblasts from both subjects. K Sarafoglou provided clinical consultation on endocrinology issues and assisted with writing of the manuscript section dealing with clinical presentation and some tables. JL Fridovich-Keil oversaw the project, wrote some sections of the manuscript, and edited and finalized all sections of the manuscript for submission. All authors assisted with editing the final manuscript.

Guarantor

Judith L. Fridovich-Keil is the guarantor for this work.

Competing Interest Statement

None of the authors has any competing interests to disclose.

Funding

This work was funded, in part, by grant NIH R01 DK059904 (to JLFK). "The author(s) confirm(s) independence from the sponsors; the content of the article has not been influenced by the sponsors."

Ethics Approval

This project was conducted with approval from the Emory University Institutional Review Board (Protocol # 618–99, PI: JL Fridovich-Keil) and study volunteers were consented prior to the study in accordance with that protocol.

No vertebrate animals were used in this study.

References

Adzhubei I, Schmidt S, Peshkin L et al (2010) A method and server for predicting damaging missense mutations. Nat Methods 7:248–249

Alano A, Almashanu S, Chinsky JM et al (1998) Molecular characterization of a unique patient with epimerase-deficiency galactosaemia. J Inher Metab Dis 21:341–350

Bang Y, Nguyen T, Trinh T, Kim Y, Song J, Song Y (2009) Functional analysis of mutations in UDP-galactose-4-epimerase (GALE) associated with galactosemia in Korean patients using mammalian GALE-null cells. FEBS J 276:1952–1961

Chhay J, Openo K, Eaton J, Gentile M, Fridovich-Keil J (2008a) A yeast model reveals biochemical severity associated with each of three variant alleles of galactose-1P uridylyltransferase segregating in a single family. J Inherit Metab Dis 31:97–107

Chhay J, Vargas C, McCorvie T, Fridovich-Keil J, Timson D (2008b) Analysis of UDP-galactose 4'-epimerase mutations associated with the intermediate form of type III galactosaemia. J Inherit Metab Dis 31:108–116

Frey PA (1996) The Leloir pathway: a mechanistic imperative for three enzymes to change the sterochemical configuration of a single carbon in galactose. The FASEB 10:461–470

Fridovich-Keil JL, Walter JH (2008) Galactosemia. In: D Valle, A Beaudet, B Vogelstein, K Kinzler, S Antonarakis, and A Ballabio (eds) The online metabolic & molecular bases of inherited disease. McGraw Hill, Columbus, OH http://www.ommbid.com/

Fridovich-Keil JL, Bean L, He M, Schroer R (2011) Epimerase Deficiency Galactosemia. In: RA Pagon, TD Bird, CR Dolan, K Stephens, and MP Adam (eds) GeneReviews™ [Internet]. Seattle (WA): University of Washington, Seattle; 1993

Gitzelmann R, Steimann B (1973) Uridine diphosphate galactose 4-epimerase deficiency. Helv Paediat Acta 28:497–510

Gitzelmann R, Steinmann B, Mitchell B, Haigis E (1976) Uridine diphosphate galactose 4'-epimerase deficiency. Helv Paediat Acta 31:441–452

Henderson JM, Huguenin SM, Cowan TM, Fridovich-Keil JL (2001) A PCR-based method for detecting known mutations in the human UDP galactose-4'-epimerase gene associated with epimerase-deficiency galactosemia. Clin Genetics 60(5):350–355

Henderson MJ, Holton JB (1983) Further observations in a case of uridine diphosphate galactose-4-epimerase deficiency with a severe clinical presentation. J Inher Metab Dis 6:17–20

Holton JB, Gillett MG, MacFaul R, Young R (1981) Galactosemia: a new severe variant due to uridine diphosphate galactose-4-epimerase deficiency. Arch Dis Child 56:885–887

Maceratesi P, Dallapiccola B, Novelli G, Okano Y, Isshiki G, Reichardt JKV (1996) Missense mutations in Japanese patients with epimerase deficiency galactosemia. Am J Hum Gen 59((Supplement)):A204

Maceratesi P, Daude N, Dallapiccola B et al (1998) Human UDP-Galactose 4' Epimerase (GALE) gene and identification of five missense mutations in patients with epimerase-deficiency galactosemia. Molecular Genetics Metabolism 63:26–30

Mitchell B, Haigis E, Steinmann B, Gitzelmann R (1975) Reversal of UDP-galactose 4-epimerase deficiency of human leukocytes in culture. Proc Nat Acad Sci 72:5026–5030

Neitzel H (1986) A routine method for the establishment of permanent growing lymphoblastoid cell lines. Hum Genet 73:320–326

Ng P, Henikoff S (2003) SIFT: predicting amino acid changes that affect protein function. Nucleic Acids Res 31:3812–3814

Ono H, Mawatari H, Mizoguchi N, Eguchi T, Sakura N, Hamakawa M (1999) Transient galactosemia detected by neonatal mass screening. Pediatr Int 41:281–284

Openo K, Schulz J, Vargas C et al (2006) Epimerase-deficiency galactosemia is not a binary condition. Am J Hum Genet 78:89–102

Park H, Park K, Kim J et al (2005) The molecular basis of UDP-galactose-4-epimerase (GALE) deficiency galactosemia in Korean patients. Genet Med 7:646–649

Quimby BB, Alano A, Almashanu S, DeSandro AM, Cowan TM, Fridovich-Keil JL (1997) Characterization of two mutations associated with epimerase-deficiency galactosemia using a yeast expression system for human UDP-galactose-4-epimerase. Am J Hum Gen 61:590–598

Ross KL, Davis CN, Fridovich-Keil JL (2004) Differential roles of the Leloir pathway enzymes and metabolites in defining galactose sensitivity in yeast. Mol Gen Metab 83:103–116

Sanders R, Sefton J, Moberg K, Fridovich-Keil J (2010) UDP-galactose 4' epimerase (GALE) is essential for development of Drosophila melanogaster. Dis Model Mech 3(9–10):628–638

Sardharwalla IB, Wraith JE, Bridge C, Fowler B, Roberts SA (1988) A patient with severe type of epimerase deficiency galactosemia. J Inher Metab Dis 11:249–251

Sarkar M, Bose S, Mondal G, Chatterjee S (2010) Generalized epimerase deficiency galactosemia. Indian J Pediatr 77(8):909–910

Shin Y (1991) In: Holmes F (ed) Techniques in diagnostic human biochemical genetics: a laboratory manual. Wiley-Liss, New York, pp 267–283

Thoden JB, Frey PA, Holden HM (1996) Molecular structure of the NADH/UDP-glucose abortive complex of UDP-galactose 4-epimerase from Escherichia coli: implications for the catalytic mechanism. Biochemistry 35(16):5137–5144

Thoden JB, Wohlers TM, Fridovich-Keil JL, Holden HM (2001a) Human UDP-galactose 4-epimerase: accommodation of UDP-N-acetylglucosamine within the active site. J Biol Chem 276:15131–15136

Thoden JB, Wohlers TM, Fridovich-Keil JL, Holden HM (2001b) Molecular basis for severe epimerase-deficiency galactosemia: X-ray structure of the human V94M-substituted UDP-galactose 4-epimerase. J Biol Chem 276:20617–20623

Thomas P, Campbell M, Kejariwal A et al (2003) PANTHER: a library of protein families and subfamilies indexed by function. Genome Res 13:2129–2141

Timson D (2005) Functional analysis of disease-causing mutations in human UDP-galactose 4-epimerase. FEBS J 272:6170–6177

Walter JH, Roberts REP, Besley GTN et al (1999) Generalised uridine diphosphate galactose-4-epimerase deficiency. Arch Dis Child 80:374–376

Wasilenko J, Fridovich-Keil J (2006) Relationship between UDP galactose 4'-epimerase activity and galactose sensitivity in yeast. J Biol Chem 281:8443–8449

Wohlers T, Fridovich-Keil JL (2000) Studies of the V94M-substituted human UDP-galactose-4-epimerase enzyme associated with generalized epimerase-deficiency galactosemia. J Inher Metab Dis 23:713–729

Wohlers TM, Christacos NC, Harreman MT, Fridovich-Keil JL (1999) Identification and characterization of a mutation, in the human UDP galactose-4-epimerase gene. Associated with generalized epimerase-deficiency galactosemia. Am J Hum Gen 64:462–470

JIMD Reports
DOI 10.1007/8904_2012_154

RESEARCH REPORT

Lyso-Gb3 Indicates that the Alpha-Galactosidase A Mutation D313Y is not Clinically Relevant for Fabry Disease

Markus Niemann • Arndt Rolfs • Anne Giese • Hermann Mascher • Frank Breunig • Georg Ertl • Christoph Wanner • Frank Weidemann

Received: 14 December 2011 / Revised: 27 April 2012 / Accepted: 14 May 2012 / Published online: 1 July 2012
© SSIEM and Springer-Verlag Berlin Heidelberg 2012

Abstract The X-chromosomal-linked lysosomal storage disorder Fabry disease can lead to life-threatening manifestations. The pathological significance of the Fabry mutation D313Y is doubted, because, in general, D313Y patients do not present clinical manifestations conformable with Fabry disease. This is in contrast to the analysis of the alpha-galactosidase A activity, which is reduced in D313Y patients. We report a comprehensive clinical, biochemical and molecular genetic analysis of two patients with a D313Y mutation. The alpha-galactosidase A activity was reduced in both patients. No Fabry symptoms or Fabry organ involvement was detected in these patients. The new biomarker lyso-Gb3, severely increased in classical Fabry patients, was determined and in both patients lyso-Gb3 was below the average of a normal population.

Our data for the first time not only clinically but also biochemically supports the hypothesis that the D313Y mutation is not a classical one, but a rare variant mutation.

Communicated by: Frits Wijburg

Competing interests: none declared

Authors Markus Niemann and Arndt Rolfs contributed equally

M. Niemann · F. Breunig · G. Ertl · C. Wanner · F. Weidemann
Department of Internal Medicine I, University of Würzburg, Würzburg, Germany

M. Niemann · F. Breunig · G. Ertl · C. Wanner · F. Weidemann
Comprehensive Heart Failure Center, University of Würzburg, Würzburg, Germany

A. Rolfs · A. Giese
Albrecht-Kossel Institute for Neuroregeneration, University of Rostock, Rostock, Germany

H. Mascher
Pharm-analyt Laboratory, Analytical department, Baden, Austria

F. Weidemann (✉)
Medizinische Klinik und Poliklinik I, Oberdürrbacher Str. 6, 97080, Würzburg, Germany
e-mail: Weidemann_F@medizin.uni-wuerzburg.de

Report

Fabry disease is an X-chromosomal-linked lysosomal storage disorder caused by decreased activity of the enzyme alpha-galactosidase A (Desnick et al. 1995). Classical Fabry disease mainly affects three organs: the kidney, the heart and the central nervous system (Whybra et al. 2009). This can lead to life-threatening manifestations at end stage of the disease including severe cardiomyopathy with arrhythmias, renal failure and stroke (Whybra et al. 2009). Besides the classical form of the disease, in recent years, so-called Fabry organ-variants have been described which mainly affect only the heart or the kidney. Moreover, some reports suggest that also non-organ-affecting variants, which do not alter organ function, may exist (Linthorst et al. 2010; Houge et al. 2011). These non-organ-affecting variants tend to have a higher frequency in the normal population than the classical private Fabry mutations, which are often only found in one single family (Linthorst et al. 2010). The variant mutation D313Y (located on an exon) seems to be such a mutation. The allele frequency in the normal population of D313Y is approximated as high as 0.5 % (Yasuda et al. 2003). Initially the D313Y mutation was described as causing classical Fabry disease by Eng et al. in 1993 (Eng et al. 1993), although Eng et al. did not include extensive organ characterisation in their study. Newer reports doubted the pathological significance of this special mutation, even the possibility that D313Y is a polymorphism was discussed (Froissart et al. 2003; Yasuda et al. 2003). In some of the studies, the alpha-galactosidase A activity in patients with the D313Y mutation was decreased, in others it was not (Froissart et al. 2003; Yasuda et al. 2003; Baptista et al. 2010; Gaspar et al. 2010; Wozniak et al. 2010). Because of the phenotypic and biochemical partly discrepant results, a large Portuguese stroke study screening for Fabry disease comes to the conclusion, however, that the pathogenicity of the mutation D313Y cannot be conclusively determined without additional information (Baptista et al. 2010).

We report a comprehensive clinical, biochemical and molecular genetic analysis of two patients with a pure D313Y mutation (out of our cohort of 175 Fabry patients). The study conformed to the principles outlined in the Declaration of Helsinki and the locally appointed ethics committee has approved the research protocol. The referral to our Fabry centre took place because of the query for initiation of enzyme replacement therapy in a 20-year-old female patient with proven D313Y mutation. The patient had been screened for Fabry disease in another university hospital because of skin lesions that were found on the arms and legs, but also on the trunk and face. Her only clinical symptom was unspecific pain manifested in her arms. A reduced alpha-galactosidase A activity of 0.35 nmol/min/mg protein (normal: 0.4–1.0 nmol/min/mg protein) and the described D313Y mutation were detected. Subsequently, an extensive family pedigree and testing of first-grade relatives was performed. The results showed no mutation in the mother's gene with a normal alpha-galactosidase A activity with 0.52 nmol/min/mg protein. The father, however, had a D313Y mutation and a slightly reduced alpha-galactosidase activity of 0.32 nmol/min/mg protein. We conducted a detailed clinical analysis of the two family members with proven mutations in our centre (Table 1). Neither the 20-year-old female nor her father (53 years old) showed any signs or organ manifestations linked to classical Fabry manifestation. The only clinical diagnosis which could be made in the father was a long-lasting and excessive nicotine abuse. In the young female, a new dermatological examination with skin biopsies gave evidence for the presence of keratosis pilaris rubra atrophicans, but no evidence of typical Fabry angioma. In addition to alpha-galactosidase A activity and molecular testing, we determined lyso-Gb3 in these two patients. For lyso-Gb3, lyso-Ceramide had been used as reference items (Matreya LLC, Pleasant Gap, PA, USA) and D5-Fluticasone Propionate (EJY Tech, Inc., Rockville, MD, USA) were used as internal standards. The method was performed analogous to the method published in Tanislav et al. (2011). The lyso-Gb3 in the daughter was measured 0.22 ng/ml, which is below the average of a normal population (around 0.4 ng/ml) (Tanislav et al. 2011). No lyso-Gb3 was detectable in her father's blood.

Table 1 Clinical characteristics of the two Fabry patients with the D313Y mutation

	Daughter	Father
Age (years)	20	53
Alpha-galactosidase activity (nmol/min/ mg protein)	0.35	0.32
LysoGb3 (ng/ml)	0.22	0
BMI (kg/m²)	25.0	22.1
Heart rate (/min)	60	62
Blood pressure (mmHg)	115/65	116/82
Cardiac assessment		
LVED wall-thickness (mm)	7	8
Left ventricular mass (g/m²) (Devereux formula)	42	62
EF (%)	69	69
Diastolic function	Normal	Normal
Strain rate lateral wall (s⁻¹)*	1.3	1.2
MRI	No LE	No LE
Oedema	No	No
Renal assessment		
Creatinin plasma (mg/dl)	0.7	0.8
DTPA-clearance (ml/min)	123	124
Proteinuria	No	No
Albuminuria	No	No
Neurological assessment		
Stroke	No	No
Transient ischaemic attack	No	No
Pain	Yes, atypical	No
Acroparaesthesia	No	No
Symptoms		
Abnormal sweating	No	No
Heat or cold intolerance	No	No
Sudden deafness	No	No
Angiokeratomata	0	0
Dyspnoea on exertion	No	No

BMI body mass index, *DTPA* diethylene triamine pentaacetic acid, *EF* ejection fraction, *Gb3* globotriaosylceramide, *GFR* glomerular filtration rate, *LE* late enhancement, *LVED* left ventricular end-diastole, *MRI* magnetic resonance tomography

*Normal absolute values for systolic strain rate in the lateral wall (regional myocardial function) > 1.1

The detailed analysis of our two patients provides clear evidence that the mutation D313Y causes a pseudodeficiency of the alpha-galactosidase A, not associated with Fabry disease. Even in the 53-year-old father (an age at which male Fabry patients often die), no Fabry-associated manifestations could be determined. Thus, our data challenge the usual way to establish the diagnosis in Fabry disease (Havndrup et al. 2010; Weidemann and Niemann 2010): When a male patient shows decreased alpha-galactosidase A activity (like our male patient), the diagnosis is regarded as proven. In female patients with borderline alpha-galactosidase A activity (like in our female patient), genotyping with the search for a Fabry-related mutation is demanded (Weidemann and Niemann 2010). However, these two conditions are perfectly met in our patients with the D313Y mutation. Thus, our female and male patients are very good examples that the pure assessment of the alpha-galactosidase A activity in combination with genotyping is not sufficient for diagnosing Fabry disease (which is especially a problem for the future in screening studies (Linthorst et al. 2010; Houge et al. 2011). In contrast, lyso-Gb3 seems to be the biochemical key for the clinical classification of an unclear alpha-galactosidase A mutation. In 2008, the new bio-marker lyso-Gb3, a degradation product of the stored lyso-Gb3, was proposed for the first time to be a hallmark of Fabry disease by Aerts et al. (2008). They could show that lyso-Gb3 was elevated in patients with classical Fabry disease. Although the pathophysiological role of lyso-Gb3 in Fabry disease is not elucidated and mechanisms in Fabry organ complications involve multiple mechanisms beyond lyso-Gb3 (Schiffmann 2009; Auray-Blais et al. 2010; Brakch et al. 2010), the results of the study by Aerts et al. suggested that lyso-Gb3 plays an important role as a factor in the pathogenesis and progression of the disease (Aerts et al. 2008). Moreover, the same group also demonstrated little to normal values in atypical mutations (van Breemen et al. 2011). In our two patients, only a very low or no lyso-Gb3 at all was detectable, indicating that there was no classic organ involvement in these patients – as proven by our clinical characterisation. The explanations for this might be as follows: (1) In general, mutated and often misfolded alpha-galactosidase A proteins do not reach the lysosomes but are stuck in the endoplasmatic reticulum. However, there is evidence that the D313Y mutated alpha galactosidase A proteins do reach the lysosomes (Yasuda et al. 2003). (2) In addition, a normal activity of the D313Y-mutated protein was found in the lysosomes and a pseudo-reduced activity in plasma by Yasuda et al. (2003). This is due to the pH dependency of the D313Y-mutated alpha galactosidase A protein which results in a normal activity in the acid lysosomes and a reduced activity in the neutral plasma. (3) The mutation is located at some distance from the active site of the protein and the dimer interface. (4) The asparagine 313 has expressed its carboxyl-end to the exterior (hydrophilic fraction), while the rest of the side chain is in a hydrophobic environment (Froissart et al. 2003; Yasuda et al. 2003). Therefore, Yasuda et al. concluded that an exchange from asparagine to tyrosine could be tolerated without greater damage for protein activity (Yasuda et al. 2003). This all leads to a residual high activity of alpha-galactosidase A in the lysosomes leading to only small amounts of Gb3 storage. Because

Lyso-Gb3 is a degradation product of Gb3 the residual activity of alpha-galactosidase A also leads to only a small amount of lyso-Gb3, which indicates a non-organ-affecting variant.

In conclusion, our data strengthen the hypothesis that the D313Y mutation is not a classical phenotype mutation but a rare variant mutation located on an exon as described by Froissart, Yasuda and Desnick. For the first time this assumption can also be supported biochemically by the measurement of the new marker lyso-Gb3. There are several other mutations (e.g. R112H, c.593C4T) where a discrepancy between positive standard diagnostic tests and lack of manifest disease have been shown (Aerts et al. 2008; Houge et al. 2011). It is a task for the future to perform lyso-Gb3 and other biochemical testing in various Fabry mutations, which are described as atypical in the literature. Of course, we cannot deny – due to the small number of our patients' cohort, the main limitation of our study – that there might be some patients with D313Y, where the biochemical analysis (enzyme activity and lyso-Gb3) might argue for a mild phenotype. This has to be discussed since it might be possible that the clinical spectrum of resulting consequences in some mutations reflects a continuum instead of a yes/no spectrum. The biochemical analysis of Fabry patients is even more complicated by the missing standardisation of measurements of the biomarkers involved in the diagnosis of Fabry disease. The standardisation is a major task for the future.

Limitations: The small number of our patient's cohort, leading to the diagnostic consequences explained in the last paragraph of the discussion.

Take Home Message

Our Lyso-Gb3 analysis indicates that the alpha-galactosidase A mutation D313Y is not clinically relevant for Fabry disease and questions the usual way to diagnose Fabry disease.

Conflict of Interest

None declared.

References

Aerts JM, Groener JE, Kuiper S et al (2008) Elevated globotriaosyl-sphingosine is a hallmark of Fabry disease. Proc Natl Acad Sci USA 105:2812–2817

Auray-Blais C, Ntwari A, Clarke JT et al (2010) How well does urinary lyso-Gb3 function as a biomarker in Fabry disease? Clinica chimica acta; Int J Clin Chem 411:1906–1914

Baptista MV, Ferreira S, Pinho EMT et al (2010) Mutations of the GLA gene in young patients with stroke: the PORTYSTROKE study–screening genetic conditions in Portuguese young stroke patients. Stroke 41:431–436

Brakch N, Dormond O, Bekri S et al (2010) Evidence for a role of sphingosine-1 phosphate in cardiovascular remodelling in Fabry disease. Eur Heart J 31:67–76

Desnick R, Ionnou Y, Eng C (1995) Fabry disease: alpha galactosidase A deficiency. In: Scriver C, Beaudet A, Sly W, Valle D (eds) The metabolic and molecular bases of inherited disease. McGraw Hill, New York, pp 2741–2784

Eng CM, Resnick-Silverman LA, Niehaus DJ, Astrin KH, Desnick RJ (1993) Nature and frequency of mutations in the alpha-galactosidase A gene that cause Fabry disease. Am J Hum Genet 53:1186–1197

Froissart R, Guffon N, Vanier MT, Desnick RJ, Maire I (2003) Fabry disease: D313Y is an alpha-galactosidase A sequence variant that causes pseudodeficient activity in plasma. Mol Genet Metab 80:307–314

Gaspar P, Herrera J, Rodrigues D et al (2010) Frequency of Fabry disease in male and female haemodialysis patients in Spain. BMC Med Genet 11:19

Havndrup O, Christiansen M, Stoevring B et al (2010) Fabry disease mimicking hypertrophic cardiomyopathy: genetic screening needed for establishing the diagnosis in women. Eur J Heart Fail 12:535–540

Houge G, Tondel C, Kaarboe O, Hirth A, Bostad L, Svarstad E (2011) Fabry or not Fabry–a question of ascertainment. Eur J Human Genetics: EJHG 19:1111

Linthorst GE, Bouwman MG, Wijburg FA, Aerts JM, Poorthuis BJ, Hollak CE (2010) Screening for Fabry disease in high-risk populations: a systematic review. J Med Genet 47:217–222

Schiffmann R (2009) Fabry disease. Pharmacol Ther 122:65–77

Tanislav C, Kaps M, Rolfs A et al (2011) Frequency of Fabry disease in patients with small-fibre neuropathy of unknown aetiology: a pilot study. Eur J Neurol 18:631–636

van Breemen MJ, Rombach SM, Dekker N et al (2011) Reduction of elevated plasma globotriaosylsphingosine in patients with classic Fabry disease following enzyme replacement therapy. Biochim Biophys Acta 1812:70–76

Weidemann F, Niemann M (2010) Screening for Fabry disease using genetic testing. Eur J Heart Fail 12:530–531

Whybra C, Miebach E, Mengel E et al (2009) A 4-year study of the efficacy and tolerability of enzyme replacement therapy with agalsidase alfa in 36 women with Fabry disease. Genet Med 11:441–449

Wozniak MA, Kittner SJ, Tuhrim S et al (2010) Frequency of unrecognized Fabry disease among young European-American and African-American men with first ischemic stroke. Stroke 41:78–81

Yasuda M, Shabbeer J, Benson SD, Maire I, Burnett RM, Desnick RJ (2003) Fabry disease: characterization of alpha-galactosidase A double mutations and the D313Y plasma enzyme pseudodeficiency allele. Hum Mutat 22:486–492

JIMD Reports
DOI 10.1007/8904_2012_156

RESEARCH REPORT

High Incidence of Symptomatic Hyperammonemia in Children with Acute Lymphoblastic Leukemia Receiving Pegylated Asparaginase

Katja MJ Heitink-Pollé · Berthil H.C.M.T. Prinsen ·
Tom J de Koning · Peter M van Hasselt ·
Marc B Bierings

Received: 27 February 2012 / Revised: 14 May 2012 / Accepted: 16 May 2012 / Published online: 1 July 2012
© SSIEM and Springer-Verlag Berlin Heidelberg 2012

Abstract Asparaginase is a mainstay of treatment of childhood acute lymphoblastic leukemia. Pegylation of asparaginase extends its biological half-life and has been introduced in the newest treatment protocols aiming to further increase treatment success. Hyperammonemia is a recognized side effect of asparaginase treatment, but little is known about its incidence and clinical relevance. Alerted by a patient with severe hyperammonemia after introduction of the new acute lymphoblastic leukemia protocol, we analyzed blood samples and clinical data of eight consecutive patients receiving pegylated asparaginase (PEG-asparaginase) during their treatment of acute lymphoblastic leukemia. All patients showed hyperammonemia (>50 μmol/L) and seven patients (88 %) showed ammonia concentrations > 100 μmol/L. Maximum ammonia concentrations ranged from 89 to 400 μmol/L. Symptoms varied from mild anorexia and nausea to headache, vomiting, dizziness, and lethargy and led to early interruption of PEG-asparaginase in three

Communicated by: Claude Bachmann

Both last authors contributed equally

K.M.J. Heitink-Pollé (✉) · M.B. Bierings
Department of Pediatric Hematology-Oncology, University
Medical Center Utrecht/Wilhelmina Children's Hospital, Room
number KC 03.063.0, Postbox 85090, 3508 AB Utrecht,
The Netherlands
e-mail: kheitink@umcutrecht.nl

B. H.C.M.T. Prinsen
Department of Metabolic Diseases, University Medical Center
Utrecht/Wilhelmina Children's Hospital, Utrecht, The Netherlands

T.J. de Koning · P.M. van Hasselt
Department of Pediatric Metabolic Disease, University Medical
Center Utrecht/Wilhelmina Children's Hospital, Utrecht,
The Netherlands

patients. No evidence of urea cycle malfunction was found, so overproduction of ammonia through hydrolysis of plasma asparagine and glutamine seems to be the main cause. Interestingly, ammonia concentrations correlated with triglyceride values ($r = 0.68$, $p < 0.0001$), suggesting increased overall toxicity.

The prolonged half-life of PEG-asparaginase may be responsible for the high incidence of hyperammonemia and warrants future studies to define optimal dosing schedules based on ammonia concentrations and individual asparagine and asparaginase measurements.

Introduction

In the past four decades, overall survival of childhood acute lymphoblastic leukemia (ALL) has improved dramatically from 34 % in the early 1970s to 86 % in the year 2004 (Kamps et al. 2010). Addition of asparaginase to treatment protocols has contributed significantly to this improved outcome (Muller and Boos 1998).

Asparaginase depletes plasma of the nonessential amino acid asparagine by hydrolyzing it into aspartic acid and ammonia. Since leukemic cells possess insufficient asparagine synthetase activity, an intracellular deficiency of asparagine leads to inhibition of protein synthesis and subsequent cell death (Prager and Bachynsky 1968).

Three different preparations of asparaginase are used in current ALL treatment protocols: native *Escherichia coli* L-asparaginase, *Erwinia chrysanthemi* L-asparaginase, and pegylated asparaginase (PEG-asparaginase). Pegylation of L-asparaginase reduces immunogenic potential and extends the half-life of the enzyme activity from 1.3 days to 5.7 days (Asselin et al. 1993).

DCOG - INTENSIFICATION/CONTINUATION MR PATIENTS part 1

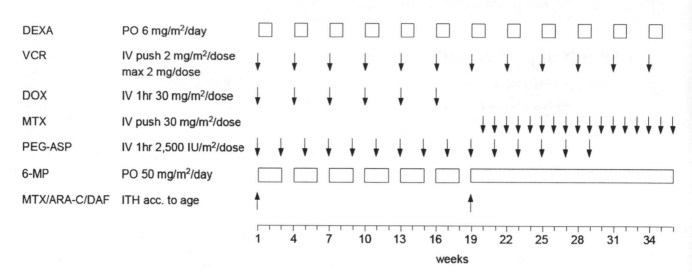

Fig. 1 Schematic drawing of the Dutch Childhood Oncology Group intensification/continuation of medium-risk patients. This part of therapy starts 20 weeks after diagnosis. *DEXA* dexamethasone, *VCR* vincristine, *DOX* doxorubicine, *MTX* methotrexate, *PEG-ASP* PEG-asparaginase; *6-MP* 6 mercaptopurine, *ARA-C* cytarabine, *DAF* di-adreson-F, *PO* per os, *IV* intravenously, *push* rapid injection; *1 h* infusion during 1 h, *ITH* intrathecal

Despite its proven efficacy, several side effects of asparaginase treatment have been reported: hypersensitivity reactions, disturbed liver functions, coagulation disorders, pancreatitis, hyperlipidemia, and silent inactivation by antibodies (Muller and Boos 1998).

Hyperammonemia due to asparaginase therapy has been described in a number of case reports (Leonard and Kay 1986; Alvarez and Zimmerman 2000; Laterza et al. 2003; Jorck et al. 2011), but little is known about its incidence, causative mechanism, and clinical relevance. The influence of the prolonged half-life of PEG-asparaginase on the duration and severity of hyperammonemia is unknown as well. Therefore, we studied ammonia concentrations and other metabolic parameters in pediatric ALL patients receiving PEG-asparaginase and observed hyperammonemia (>50 μmol/L) in all patients and ammonia concentrations >100 μmol/L in seven out of eight patients.

Design and Methods

After presentation of the index case (patient A), we analyzed ammonia concentrations in seven consecutive patients treated in our hospital and assigned to the medium risk group of the Dutch Childhood Oncology Group (DCOG) ALL 10 protocol during their treatment with PEG-asparaginase. Measurements were conducted between 2007 and 2009.

The DCOG ALL 10 protocol is a protocol based on the pediatric ALL protocol of the German Berlin-Frankfurt-

Münster (BFM) Group, consisting of induction, high-dose methotrexate, followed by minimal residual disease–based assignment to standard, medium, or high-risk intensification and continuation treatment. The outline of the intensification/continuation treatment for medium-risk-group patients is shown in Fig. 1.

Ammonia concentrations were analyzed before the first administration of PEG-asparaginase, at least once, 1 week after PEG-asparaginase when also dexamethasone was administered in the previous week (week 2, 8, 14, 20, or 26) as well as 1 week after PEG-administration when no dexamethasone was administered (week 4, 6, 10, 12, 16, 18, 22, 24, 28, or 30). Hyperammonemia was defined as ammonia concentrations >50 μmol/L and clinical significant hyperammonemia as ammonia concentrations >100 μmol/L (Laterza et al. 2003; Cohn and Roth 2004; Steiner et al. 2007). At the above-mentioned time points, we also measured liver enzymes (alanine transaminase, aspartate transaminase, gamma glutamyl transferase, and alkaline phosphatase), bilirubin, total amylase, lipase, triglycerides, and cholesterol. In case of elevated ammonia concentrations or symptoms or signs suggestive of hyperammonemia, repeated ammonia measurements were performed. Blood samples were transported to the lab, centrifuged in the cold, and immediately stored at −20°C to deactivate the effect of PEG-asparaginase. Blood samples for ammonia analysis were generally taken 3–4 h after the last meal and analyzed by standardized enzymatic-photometric assay using a DxC 800 analyzer (Beckman Coulter diagnostics division, Brea, CA, USA).

To rule out secondary abnormalities such as urea cycle disorders, organic acidurias or fatty acid oxidation disorders, plasma amino acids, acylcarnitine concentrations, and, in six patients, urine orotic acid excretion were measured. In order to deactivate the PEG asparaginase in vitro, the blood sample tubes for amino acids and ammonia were immediately put in ice water, rapidly transported to the lab, centrifuged in the cold and the plasma deproteinized before storing the supernatant at −20°C until analysis.

Symptoms and signs associated with hyperammonemia, like headache, lethargy, nausea, vomiting, dizziness, and neurological symptoms were documented in the patient files. A waiver of the requirement for informed consent was granted by the Institutional Review Board of the University Medical Center Utrecht, which reviewed the study.

Results and Discussion

Patient characteristics and results are shown in Tables 1 and 2. Hyperammonemia was detected in all patients and ammonia concentrations >100 μmol/L in seven out of eight consecutive patients with acute ALL during treatment with PEG-asparaginase. All patients with clinically significant hyperammonemia displayed symptoms. Although generally mild, symptoms were severe in three patients (A, C, and D), leading to hospital admission and discontinuation of PEG-asparaginase therapy.

Hyperammonemia after asparaginase therapy was first reported by Leonard and Kay in 1986 (Leonard and Kay 1986), followed by a number of other cases (Alvarez and Zimmerman 2000; Laterza et al. 2003; Jorck et al. 2011). Although the exact incidence of symptomatic hyperammonemia after asparaginase therapy is unknown, the small number of cases reported suggests a low frequency. Indeed, symptomatic hyperammonemia is not mentioned in recent overviews of toxicity of asparaginase treatment (Muller and Boos 1998; Earl 2009; Rytting 2010). In contrast, our observations suggest that symptomatic hyperammonemia is the rule rather than the exception after PEG-asparaginase administration.

This discrepancy may in part be a consequence of the tendency to analyze ammonia levels only in case of severe symptoms suggestive of hyperammonemia. In support, the only study that measured ammonia concentrations regardless of symptoms observed a similar high frequency of hyperammonemia: 7 out of 10 patients described by Steiner et al. reached ammonia concentrations >100 μmol/L 1 day after L-asparaginase administration (Steiner et al. 2007). However, these patients all remained without symptoms, contrasting with the overt symptoms in our patients. One could argue that most symptoms in our patients were nonspecific and could be attributable, at least in part, to other chemotherapeutics being administered. However, patients themselves reported these symptoms to be more severe than they experienced during other parts of leukemia treatment. Moreover, all symptoms disappeared after cessation of PEG-asparaginase administration. We speculate that the higher incidence of clinical symptoms of hyperammonemia in this study is secondary to the use of PEG-asparaginase. Due to the prolonged half-life of PEG-asparaginase, ammonia concentrations may not return to normal before the next dose is administered and ammonia toxicity may accumulate. As shown in Fig. 2, patients B, D, and F show a pattern consistent with this hypothesis. The linear correlation between the number of received doses of PEG-asparaginase and ammonia concentrations (Fig. 3) also supports this hypothesis. The half-life of PEG-asparaginase is likely to be longer than the above-mentioned 5.7 days, since asparagine was still depleted 3–4 weeks after the last dose of PEG-asparaginase in three out of three patients with available data. Interestingly, ammonia concentrations correlated with triglyceride values ($r = 0.68$, $p < 0.0001$), suggesting increased overall toxicity.

Little is known about the exact mechanism through which asparaginase causes hyperammonemia. Normally, ammonia is removed rapidly from the circulation by being incorporated to glutamine in brain and muscle after which it is converted to urea via the urea cycle in the liver (Lockwood et al. 1979). Most likely, the altered equilibrium between glutamine and glutamate in the patients studied, favoring glutamate, contributes to the expansion of the ammonia pool. Given the plasma concentrations of asparagine and glutamine, the production of ammonia may well exceed the detoxification capacity in the liver. Indeed, analysis of plasma amino acid concentrations showed a complete depletion of asparagine and a marked reduction in glutamine concentration. The concentrations of aspartate (range: 21–50 μmol/L, reference values: 3–15 μmol/L, five out of six patients) and glutamate (range: 169–733 μmol/L, reference values: 17–69 μmol/L, six out of six patients) were strongly increased. Plasma concentrations of all urea cycle intermediates and excretion of orotic acid remained normal during therapy in all samples analyzed, arguing against malfunction of the urea cycle. Malnutrition during chemotherapy and proteolysis may induce catabolism, theoretically further increasing the ammonia load. However, we found no deficits of other amino acids in plasma, as would be expected in malnourished patients. Moreover, in case of weight loss during chemotherapy, tube feeding is instituted instantly in our hospital. Plasma acylcarnitine concentrations were also

Table 1 Patient characteristics and description of toxicity. *M/F* male or female, *ALL* Acute Lymphoblastic Leukemia, *PEG-asparaginase* Pegylated asparaginase, *CR* complete remission

Patient	M/F	Age at diagnosis (years)	Diagnosis	Maximum documented ammonia concentration (μmol/L)	Number of PEG-asparaginase doses before maximum ammonia concentration	Symptoms	Treatment for hyperammonemia	Other toxicities during PEG-asparaginase therapy	Outcome
A (index case)	F	5	Pre B ALL	400	6	Headache, nausea, vomiting, lethargy, leading to admission	Lactulose (1.8 mL/kg), protein restricted diet; Omission of eight doses of PEG-asparaginase	Severe hypertriglyceridemia (1,946 mg/dL), hypercholesterolemia (703 mg/dL), elevated gamma GT (2,163 U/L)	In CR, 33 months after end of therapy
B	F	3	Pre B ALL	258	13	Mild lethargy, anorexia, weakness	None	Moderate hypertriglyceridemia (690 mg/dL)	In CR, 30 months after end of therapy
C	M	3	Common ALL	248	8	Lethargy, nausea	Omission of 1 dose of PEG-asparaginase	Moderate hypertriglyceridemia (717 mg/dL) Mild hypercholesterolemia (286 mg/dL)	In CR, 28 months after end of therapy
D	F	12	Pre B ALL	281	14	Dizziness, lethargy, headache, leading to admission	Lactulose, protein restricted diet; omission of 1 dose of PEG-asparaginase	Severe hypertriglyceridemia (3,645 mg/dL), Hypercholesterolemia (792 mg/dL) thrombosis vena axillaris, elevated gamma GT (1,280 U/L), hypoglycemia (2.4 mmol/L)	In CR, 27 months after end of therapy
E	M	3	Common ALL	366	5	Mild nausea, headache, malaise	None	Severe hypertriglyceridemia (8,725 g/dL) Hypercholesterolemia (448 mg/dL)	In CR, 21 months after end of therapy
F	F	9	Pre B ALL	320	13	Mild lethargy, nausea	None	Severe hypertriglyceridemia (1,991 mg/dL) Hypercholesterolemia (471 mg/dL)	In CR, 21 months after end of therapy
G	M	5	Common ALL	123	2	Anorexia	None	None	In CR, 18 months after end of therapy
H	F	4	Common ALL	89	1	None	Not applicable	none	In CR, 17 months after end of therapy

Table 2 Results of metabolic studies in described patients. Reference values: asparagine 34–94 μmol/L; aspartate 1–17 μmol/L; glutamine 333–809 μmol/L; glutamate 14–78 μmol/L

Patient	Asparagine and aspartate	Glutamine and glutamate	Urine orotic acid
A (index case)	Asparagine: not available Aspartate elevated (21 μmol/L)	Glutamine: normal (812 μmol/L) Glutamate: not available	Normal
B	Not available	Not available	Not available
C	Asparagine: complete depletion (0 μmol/L) Aspartate: elevated (30 μmol/L)	Glutamine: decreased (325 μmol/L) Glutamate: elevated (169 μmol/L)	Normal
D	Asparagine: complete depletion (0 μmol/L) Aspartate: elevated (30 μmol/L)	Glutamine: decreased (1,295 μmol/L) Glutamate: elevated (653 μmol/L)	Normal
E	Asparagine: complete depletion (0 μmol/L) Aspartate: normal (13 μmol/L)	Glutamine: decreased (189 μmol/L) Glutamate: elevated (441 μmol/L)	Normal
F	Asparagine: complete depletion (0 μmol/L) Aspartate: elevated (24 μmol/L)	Glutamine: decreased (102 μmol/L) Gutamate: elevated (733 μmol/L)	Not available
G	Asparagine: complete depletion (0 μmol/L) Aspartate: elevated (21 μmol/L)	Glutamine: decreased (116 μmol/L) Glutamate: elevated (463 μmol/L)	Normal
H	Asparagine: complete depletion (0 μmol/L) Aspartate: normal (8 μmol/L)	Glutamine: decreased (54 μmol/L) Glutamate: elevated (263 μmol/L)	Normal

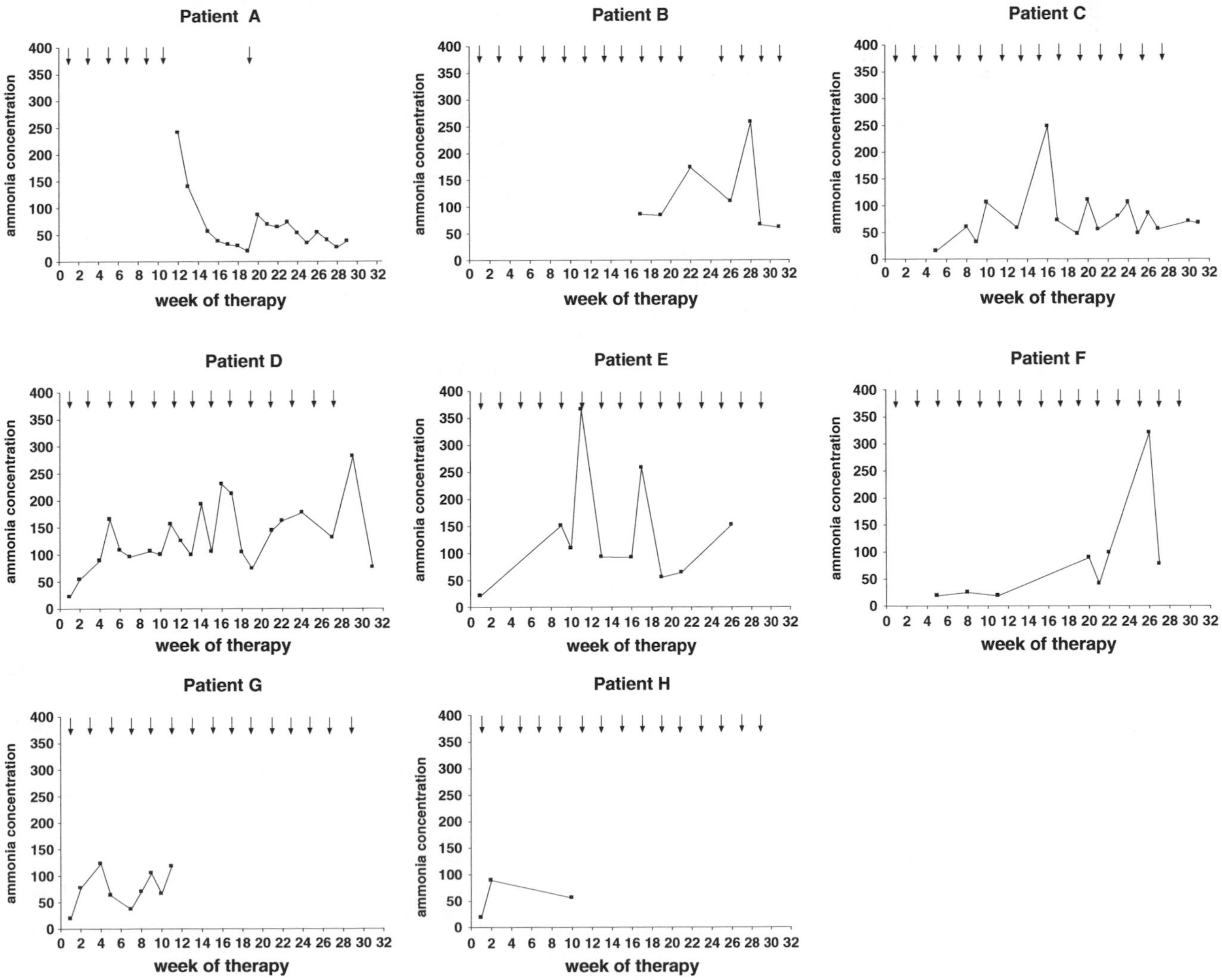

Fig. 2 Ammonia concentrations in each patient related to the week of treatment. *Arrows* denote PEG-asparaginase administration. In patient A, 8 gifts were omitted; in patients C and D, the last gift was omitted

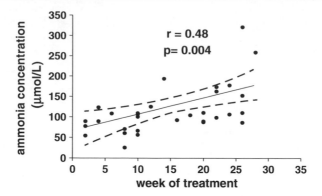

Fig. 3 Correlative analysis of ammonia concentrations 1 week after PEG-asparaginase administration and week of treatment

normal in all samples analyzed, excluding other known causes of secondary hyperammonemia.

In conclusion, our data clearly indicate that the clinical and biochemical consequences of asparaginase treatment are likely to be more pronounced when using PEG-asparaginase because of its prolonged half-life. Future studies are indicated to define the exact incidence of symptomatic hyperammonemia in patients receiving PEG-asparaginase, to define the relationship between hyperammonemia and clinical symptoms and to elucidate the link between hyperammonemia and hypertriglyceridemia. Finally, therapeutic drug monitoring by measuring ammonia concentrations, asparagine concentrations, and asparaginase levels might be of clinical use to define if lower doses of or longer intervals between PEG-asparaginase gifts are feasible. In the next DCOG ALL treatment protocol (ALL 11), therapeutic drug monitoring to optimize dosing of PEG-asparaginase will be studied.

Take Home Message

Pegylation of asparaginase, used in the treatment of childhood acute lymphoblastic leukemia, seems to result in a high incidence of symptomatic hyperammonemia due to its prolonged half-life.

Authorship and Disclosures

KH designed the research, collected and analyzed data, and wrote the chapter. BP performed and analyzed laboratory studies and revised the manuscript critically. TK revised the manuscript critically. PH designed the research, analyzed data, and wrote the chapter. MB designed the research, analyzed data, and wrote the chapter. The authors declare no conflict of interest.

References

Alvarez OA, Zimmerman G (2000) Pegaspargase-induced pancreatitis. Med Pediatr Oncol 34:200–205

Asselin BL, Whitin JC, Coppola DJ, Rupp IP, Sallan SE, Cohen HJ (1993) Comparative pharmacokinetic studies of three asparaginase preparations. J Clin Oncol 11:1780–1786

Cohn RM, Roth KS (2004) Hyperammonemia, bane of the brain. Clin Pediatr (Phila) 43:683–689

Earl M (2009) Incidence and management of asparaginase-associated adverse events in patients with acute lymphoblastic leukemia. Clin Adv Hematol Oncol 7:600–606

Jorck C, Kiess W, Weigel JF, Mutze U, Bierbach U, Beblo S (2011) Transient hyperammonemia due to L-asparaginase therapy in children with acute lymphoblastic leukemia or non-Hodgkin lymphoma. Pediatr Hematol Oncol 28:3–9

Kamps WA, van der Pal-de Bruin KM, Veerman AJ, Fiocco M, Bierings M, Pieters R (2010) Long-term results of Dutch Childhood Oncology Group studies for children with acute lymphoblastic leukemia from 1984 to 2004. Leukemia 24:309–319

Laterza OF, Gerhardt G, Sokoll LJ (2003) Measurement of plasma ammonia is affected in patients receiving asparaginase therapy. Clin Chem 49:1710–1711

Leonard JV, Kay JD (1986) Acute encephalopathy and hyperammonaemia complicating treatment of acute lymphoblastic leukaemia with asparaginase. Lancet 1:162–163

Lockwood AH, McDonald JM, Reiman RE, Gelbard AS, Laughlin JS, Duffy TE, Plum F (1979) The dynamics of ammonia metabolism in man. Effects of liver disease and hyperammonemia. J Clin Invest 63:449–460

Muller HJ, Boos J (1998) Use of L-asparaginase in childhood ALL. Crit Rev Oncol Hematol 28:97–113

Prager MD, Bachynsky N (1968) Asparagine synthetase in normal and malignant tissues: correlation with tumor sensitivity to asparaginase. Arch Biochem Biophys 127:645–654

Rytting M (2010) Peg-asparaginase for acute lymphoblastic leukemia. Expert Opin Biol Ther 10:833–839

Steiner M, Attarbaschi A, Kastner U, Dworzak M, Haas OA, Gadner H, Mann G (2007) Distinct fluctuations of ammonia levels during asparaginase therapy for childhood acute leukemia. Pediatr Blood Cancer 49:640–642

JIMD Reports
DOI 10.1007/8904_2012_157

RESEARCH REPORT

Nutritional Consequences of Adhering to a Low Phenylalanine Diet for Late-Treated Adults with PKU

Low Phe Diet for Adults with PKU

Ingrid Wiig · Kristina Motzfeldt · Elin Bjørge Løken · Bengt Frode Kase

Received: 12 October 2011 / Revised: 14 May 2012 / Accepted: 21 May 2012 / Published online: 1 July 2012
© SSIEM and Springer-Verlag Berlin Heidelberg 2012

Abstract *Background:* The main treatment for phenylketonuria (PKU) is a low phenylalanine (Phe) diet, phenylalanine-free protein substitute and low-protein special foods. This study describes dietary composition and nutritional status in late-diagnosed adult patients adhering to a PKU diet.

Methods: Nineteen patients, followed at Oslo University Hospital in Norway, participated; median age was 48 years (range 26–66). Subjects were mild to severely mentally retarded. Food intake, clinical data and blood analyses relevant for nutritional status were assessed.

Results: Median energy intake was 2,091 kcal/day (range 1,537–3,277 kcal/day). Carbohydrates constituted 59% (range 53–70%) of the total energy, including 15% from added sugar; 26% was from fat. The total protein intake was 1.02 g/kg/day (range 0.32–1.36 g/kg/day), including 0.74 g/kg/day (range 0.13–1.07 g/kg/day) from protein substitutes. Median dietary Phe intake was 746 mg/day (range 370–1,370 mg/day). Median serum Phe was 542 µmol/L (range 146–1,310 mg/day). Fortified protein substitutes supplied the main source of micronutrients. Iron intake was 39.5 mg/day (range 24.6–57 mg/day), exceeding the upper safe intake level. Intake of folate and folic acid, calculated as dietary folate equivalents, was 1,370 µg/day (range 347–1744 µg/day), and resulted in high blood folate concentrations. Median intake of vitamin B_{12} was 7.0 µg/day (range 0.9–15.1 µg/day).

Conclusions: The diet supplied adequate protein and energy. Fortification of the protein substitutes resulted in excess intake of micronutrients. The protein substitutes may require adjustment to meet nutritional recommendations for adults with PKU.

Introduction

In Phenylketonuria (PKU; OMIM 261600), the conversion of phenylalanine (Phe) to tyrosine is restrained or blocked, due to impaired activity of the enzyme phenylalanine hydroxylase (PAH; EC 1.14.16.1) in the liver (Scriver et al. 2011). With the newborn screening programme, dietary treatment is usually started in the first days of life to avoid brain damage. The Norwegian national screening programme was instituted around 1970. Patients born prior to this date were usually diagnosed late and developed various degrees of brain damage. It has been documented that dietary treatment started on clinical indications, during childhood or in adult years, can alleviate neurological and behavioural symptoms and signs also in late diagnosed patients (Yannicelli and Ryan 1995; Baumeister and Baumeister 1998; Fitzgerald et al. 2000; Lee et al. 2009). In Norway, late diagnosed adults with PKU are offered dietary treatment on a permanent basis if they experience positive effects on neurological and behavioural symptoms during a trial period, usually lasting 3 to 6 months.

The main treatment for PKU is a low phenylalanine diet, supplemented with a Phe-free protein substitute, vitamins

Communicated by: Anita MacDonald

I. Wiig (✉) · B. F. Kase
Centre for Rare Disorders, Oslo University Hospital, P.O. Box 4950, Nydalen, N-0424 Oslo, Norway
e-mail: ingrid.wiig@ous-hf.no; bengt.frode.kase@ous-hf.no

K. Motzfeldt
Department of Pediatrics, Oslo University Hospital,
Ole Vigs gate 25, N-0366, Oslo, Norway
e-mail: kristina.motzfeldt@gmail.com

E. B. Løken
Department of Nutrition, Institute of Basic Medical Sciences, University in Oslo, P.O. Box 1046, Blindern, N-0317 Oslo, Norway
e-mail: e.b.loken@medisin.uio.no

and minerals. The aim of the diet is to reduce the level of phenylalanine in the blood and brain. Small amounts of protein from natural food products provide the essential amino acid phenylalanine. The amount of Phe tolerated and protein substitute needed vary according to the rest activity in the PAH enzyme. The amount of dietary Phe is determined by regular monitoring of Phe levels in the blood (Yi and Singh 2008; Poustie and Wildgoose 2010). To meet energy requirements, specially manufactured low protein foods such as bread, pasta and biscuits, and natural foods low in phenylalanine such as fruits, vegetables, sugar and butter, are also used.

There is little documentation regarding the nutritional consequences for adults who follow a PKU diet over years. Hence the objectives of this study were to report the nutrient intake of late-treated adults with PKU and describe dietary effects on nutritional status. We wished to investigate as to whether the diet used on a daily basis conformed to nutritional recommendations and medical treatment goals.

Methods

This was an observational cross-sectional study. Food intake was registered for 4 days, serum Phe levels and several blood biochemical parameters, relevant for assessing nutritional status were evaluated. The study was organised in connection with annual outpatient follow-up.

Subjects

The national PKU centre at Oslo University Hospital, Norway, invited 27 late-treated adult patients to participate in the study. All had adhered to diet for the last 12 months or longer. Consent was received from 21 subjects. Two patients withdrew before the study started, due to intercurrent illness and inability to record food intake. Nineteen late-treated subjects were recruited, 7 males and 12 females with a median age of 48 years (range 26–66 years). The median age at diagnosis was 3 years (range 6 months to 43 years), and median age at diet start was 27 years (range 6 months to 47 years). Subjects had bodyweight comparable to the general population in Norway; BMI showed a median 28 (range 20.2–38.5), four had BMI above 30. Before dietary treatment was instituted, participants had a median serum Phe of 1,541 μMol/L (range 1,131–2,468 μMol/L). All participants were ethnic Norwegians.

Subjects and carers received written and verbal information; all communication was undertaken by the first author. Twelve subjects, six male, six female, with a median age of 49 years (range 36–66 years) suffered severe cognitive disability and could not give informed consent. These 12 subjects lived in small, staffed care homes, and professional carers acted as informants and recorded the subjects' food intake. The remaining seven participants, one male, and six female, median age 42 years (range 26–51 years), had milder cognitive disabilities and were able to give informed consent. These subjects lived by themselves or with their parents and managed the diet with no, or limited, assistance from the community. These seven plus one of the severely disabled patients visited Oslo University Hospital for annual outpatient control, information on the study and venous blood tests. The remaining 11 participants were unable to visit the hospital, due to mental and physical disability and long distance to the hospital. Instead, the first author visited the care homes, to obtain data and give information on food recording and blood tests.

The Regional Committee for Medical Research Ethics and the Commission for Privacy Protection at Oslo University Hospital approved the study. Handling of blood samples was according to provisions in the Norwegian Biobank Act.

Food Recording

The patients or their carers kept a 4 day prospective food diary. Food intake was recorded from Wednesday to Saturday, or Sunday to Wednesday. The patients were encouraged to follow their ordinary diet during food recording. Foods were weighed on digital scales with 1 g increments and drinks were measured in decilitres. The diaries, personal recipes and wrapping paper for special products were returned by mail.

The recordings were analysed using a Norwegian commercial nutrient calculation programme "Mat pa data 4a", based on the official Norwegian table of food composition (Rimestad et al. 2001). Data for the protein substitutes and the low-protein food products were added to the database. Phenylalanine in food was calculated as 5% of the protein content or based on analysis of amino acid distribution (Weetch and MacDonald 2006). Vitamin B_{12} content was calculated manually as it was not included in the software programme. To account for different bioavailability in folate and folic acid, intake of this vitamin was calculated as Dietary folate equivalents (DFE) (Suitor and Bailey 2000). The Nordic Nutrition Recommendations were used to compare intake with recommendations and upper intake levels (Nordic Council of Ministers 2004). All participants received individual dietary advice after the study.

Blood Sampling and Analyses

Blood tests were taken the day before or during the period of food recording and after overnight fasting. For all subjects, routine tests for serum Phe were obtained by finger pricking. These were subsequently analysed by the Neonatal Screening Laboratory at Oslo University Hospital. Phe in serum was determined fluorometrically, by a method described by M.W. McCaman (McCaman and Robins 1962). Subjects had their serum Phe levels analysed routinely, four to ten times a year.

Venous blood samples were analysed at the Department of Medical Biochemistry at Oslo University Hospital. Analyses were done routinely on arrival at the laboratory and according to standard procedures: The following equipment was used for blood analysis: Amino acid profiles in serum: Amino Acid Analyser Biochrom 30, Biochrom LTD, Cambridge, UK. Serum iron analyses: Modular P800 Roche Diagnostics, Basel, Switzerland. Ferritin analyses: Modular E170 Roche Diagnostics, Basel, Switzerland. Haemoglobin analyses: Celldyn 4000, Abbott Laboratories, CA, USA. Folate and B_{12} analyses: Immulite 2000, Siemens Medical Solutions Diagnostics, DPC Cirrus Inc., Instrument Systems Division, Flanders, N.J., USA.

For two subjects, all analyses, apart from serum Phe, amino acid profiles, folate and vitamin B_{12}, were done at local hospital laboratories. For two other subjects, only capillary serum Phe was obtained, as venous blood drawing would have required general anaesthesia.

Statistics

Microsoft ® Excel 2002 SP3 was used for statistical analyses. Because of the small number of patients involved, only descriptive statistics such as median and range were used.

Results

Food and Nutrient Intake

Energy according to food groups are shown in Table 1. Intake of low-protein special foods together with protein substitutes constituted about half the consumed energy. Natural low-protein products like fruits, vegetables, sweets and soft drinks constituted approximately a fifth of the energy. The remaining energy stemmed from animal foods and miscellaneous food products like normal bread and cereals, edible fats and jams. The median intake of fruits and vegetables was 326 g (range 137–1,205 g/day). Median fibre intake was 16 g/day (range 8–38 g/day) or 1.7 g/MJ (range 0.9–3.4 g/MJ).

The median protein intake was 1.02 g/kg/day (range 0.32–1.36 g/kg/day). One subject reported a protein intake below the FAO/WHO recommendation of minimum 0.75 g/kg/day (Nordic Council of Ministers 2004). Natural protein with phenylalanine constituted about a quarter of the consumed protein. The median Phe intake was 746 mg/day (range 370–1,370 mg/day), or 10 mg/kg/day (range 6–21 mg/kg/day). The main source of Phe was food of animal origin.

Six subjects had a fat intake below the recommended minimum of 25% of energy, see Table 2. Main sources of dietary fat were margarine, oil, butter and mayonnaise, resulting in a median intake of 20 g/day (range 5–28 g/day) of polyunsaturated fatty acids (PUFA), according to recommendations for all but two subjects. All subjects ate some long-chain PUFA; 14 used commercial fish oil concentrates or cod liver oil daily, giving 0.5 g/day to 1.0 g/day of docosahexaenoic acid (DHA) and eicosapentaenoic acid (EPA) combined. Most subjects also used sandwich fillings based on mackerel or cod roe.

The amount of energy from carbohydrate was above the recommended level for eight subjects, mainly due to a high intake of added sugar. Only five subjects managed to maintain the intake of added sugar below the recommended maximum of 10% of total energy.

In order to maintain serum Phe in the therapeutic range and to meet each patient's individual protein requirement, the amounts of protein substitute taken varied. The median protein intake from protein substitute was 0.74 g/kg/day (range 0.13–1.07 g/kg/day). Five brands of protein substitute were used, see Table 3. The protein substitutes were fat free; four of them were fortified with minerals and vitamins. One subject was required to take additional supplements of vitamins, trace minerals and calcium, since the substitute used was unfortified. Use of fortified protein substitutes resulted in high intakes of most micronutrients. The highest micronutrient intakes were observed for iron, vitamin B_{12} and folic acid, see Table 1. The mean intakes of calcium, magnesium, zinc, selenium and vitamins A, D, C and remaining B-vitamins also exceeded recommendations. However, as intakes did not exceed upper safe levels of intake (Nordic Council of Ministers 2004) and blood parameters were not adversely influenced, the results are not reported here.

The main source of folic acid was the fortified protein substitutes, with additional small amounts from some low-protein special products. About 85% of DFE originated from folic acid in the protein substitutes, ranging from 174 µg/day for the subject taking Avonil to 887 µg/day for one subject taking Lophlex. As fortification is not permitted in Norwegian food products, the additional food did not contain folic acid. Dietary sources of natural folate were vegetables, fruits, and small amounts of dairy products, providing about 10% of DFE.

Table 1 Energy and nutrient intakes according to food groups in late-treated PKU patients

Food group	Energy (kcal/day)	Energy (kJ/day)	Protein (g/day)	Fat (g/day)	Carbohydrate (g/day)	Added sugar[a] (g/day)	Iron (mg/day)	DFE[b] (μg/day)	Vitamin B_{12} (μg/day)
	Median (range)								
Protein substitute	406 (84–512)	1701 (351–2,142)	58.5 (12.7–75)	0.7 (0–0.9)	45.9 (0.8–57.8)	4.2 (0–5.3)	31.7 (12.3–40)	1,156 (174–1,508)	5.4 (0.9–11.8)
Dietary food[c]	678 (401–1,385)	2,738 (1,679–5,799)	3 (1–8.5)	17.0 (8.8–34.8)	129.1 (59.1–247.7)	5.2 (0–20.1)	6.9 (0–28.7)	90 (0–230)	0
Vegetables and fruits	206 (67–524)	864 (282–2,195)	5 (1.5–12.3)	0.6 (0.1–2.1)	44.6 (13.8–120.2)	0.6 (0–6.7)	1.3 (0.4–4.4)	89 (17–251)	0
Animal foods[d]	153 (83–345)	641 (346–1,445)	8 (2.9–21)	10.6 (5.2–22.6)	4.8 (1.8–24)	0 (0–0.8)	0.9 (0.1–2.1)	10 (2–36)	1.3 (0–3.3)
Sweets[e]	239 (17–620)	999 (72–2,595)	0.3 (0–1.7)	0 (0–8.8)	58.5 (4–150.9)	52.5 (4–140)	0 (0–1.3)	0 (0–24)	0
Total intake	2,091 (1,537–3,277)	8,754 (6,435–13,717)	76.9 (32.6–106.3)	65.1 (37.2–102.4)	319.9 (211.8–494.2)	73.2 (8.5–179.2)	39.5 (24.6–57)	1,370 (347–1,744)	7.0 (0.9–15.1)

[a] All sugar, syrup or glucose syrup added by manufacturers or household cooking

[b] Folate and folic acid calculated as dietary folate equivalents (DFE)

[c] Dietary food manufactured with less protein (flour, baked goods, pasta, etc.)

[d] Food products consisting mainly of meat, fish, eggs, cows milk, yoghurt or cheese

[e] Candies, chocolate, popsicles, soft drinks

Table 2 Distribution of energy in the diet of late-treated PKU patients

	Proportion of total intake (%)	
	Median (range)	Nordic recommendations
Protein	14 (6–20)	10–20
Fat	26 (17–35)	25–35
Saturated fat	9 (5–17)	ca 10
Polyunsaturated fat	7 (2–11)	5–10
Carbohydrate	59 (53–70)	50–60
Added sugar	15 (2–31)	<10

The intake of vitamin B_{12} was lower than recommended only for the subject taking Avonil. The others had intakes according to recommendations or higher. All subjects had an iron intake greater than the recommendation. Almost all dietary iron originated from fortified protein substitutes and low protein special food. The protein substitutes contributed approximately 75% of the total iron intake. Intakes exceeded the upper intake levels of 25 mg/day for 14 subjects. The use of food containing haemic iron and natural B_{12} was minimal.

Blood Tests

Median serum Phe at the time of the study was 542 μMol/L (range 146–1,310 μMol/L). The subjects' mean serum Phe level during the 12 months preceding the study showed a median of 472 μMol/L (range 352–1,143 μMol/L). Phe levels at the time of the study and for the preceding year had a Spearman's rho correlation of 0.83 ($p < 0.01$). Other blood tests were analysed for 13 to 17 subjects. Folate, B_{12}, and iron parameters are shown in Table 4. Data for renal function, lipids, zinc, selenium and magnesium were within normal ranges and are not reported. Neither did the serum amino acid profiles reveal discrepancies apart from elevated serum Phe.

Discussion

We wished to describe nutritional implications for late-diagnosed adults who had followed a Phe restricted diet for at least 1 year preceding the study. PKU is a rare disease, and even when all eligible patients in Norway were invited, the sample size was small. The medical ethical committee required that the burden of participation should be minimal for retarded subjects. Thus, only blood tests commonly done at annual follow-up were allowed. However, all subjects had capillary serum Phe levels analysed routinely, and the stable Phe levels indicated that the diet was well adhered to. Most subjects chose to maintain serum Phe above the Norwegian treatment goal of maximum 400 μMol/L. For these adults, the optimal serum Phe level was individually chosen and can be seen as a compromise between the effects on patients' emotional and behavioural functions and their ability to adhere to and manage the diet. The Phe levels in this study are comparable to the levels reported in other studies for late-diagnosed patients with PKU (Lee et al. 2009; Trefz et al. 2011) and to the levels reported to have positive effects on early-treated adult patients' mood and attention (Ten Hoedt et al. 2011).

The stable serum Phe levels and the relatively high energy intakes reported support our assumption that the food recordings reflected patients' habitual dietary intake. The detailed and meticulously recorded registrations supplied important data on how the patients chose to eat at home.

Table 3 The contents of selected nutrients in the different types of protein substitutes used by late-treated PKU patients

	Type of protein substitutes				
	XP Maxamum[a]	Lophlex powder[a]	PKU Express[b]	Avonil[c]	Prekunil[c]
Number of subjects using each substitute	13	1	3	1	1
Nutrientsper 100 g of substitute					
Protein equivalents (g)	30	72	60	55.8	55
Iron (mg)	23.5	19.2	21.6	16	0
Vitamin B_{12} (μg)	3.6	6.4	9.4	1.1	0
Folic acid (μg)	500	896	400	226	0
DFE[d] (μg)	850	1,523	680	384	0

[a] XP Maxamum and Lophlex powder by Nutricia

[b] PKU Express by Vitaflo International Ltd

[c] Avonil and Prekunil by Prekulab Ltd

[d] Dietary Folate Equivalents (DFE): 1 μg folic acid as a fortificant = 1.7 μg DFE

Table 4 Nutritional biochemical parameters of the late-treated PKU patients

	Median (range)	Normal range at the laboratory used
B-Haemoglobin ($n = 13$), g/dL	13.5 (12.9–15.2)	Men: 13.4–17 Women: 11.7–17
S-iron ($n = 15$), g/L	17 (7–27)	9–34
P-Ferritin ($n = 16$), µg/L	53 (9–257)	Men: 29–383 Women: 10–167
S-Vitamin B12 ($n = 17$), pmol/L	580 (110–1,030)	160–710
ER-folate ($n = 13$), nmol/L	1,870 (605–3,165)	390–1,140
S-folate ($n = 15$), nmol/L	53.0 (22.3–>54.4)	7.1–27

N number of subjects tested

Such knowledge is of great value for dieticians and doctors when giving dietary advice and prescribing medical food.

As reported by MacDonald et al. (2003), fruits and vegetables can be used almost without restriction in the PKU diet. However, several subjects in the study consumed only small amounts of these foods, thus restricting their diet more than necessary. Intake of fruits and vegetables was comparable to the Norwegian mean intake, and might reflect what patients and carers considered normal portions. Availability and food price might also influence the use of these foods. Fruits and vegetables were the main source of fibre in the PKU diet. Only individuals with high intakes managed to reach the recommended intake of 25–35 g fibre per day. Readily available, low-cost sweets and soft drinks were used in considerable amounts.

A high sugar intake combined with low fat content in most low-protein special food used might increase the risk of deficiency of essential fatty acids, and, indeed, low intakes of PUFA in PKU diets are reported in several studies. Most studies, like the study by Rose et al., are performed on children (Rose et al. 2005). Mosely et al. reported that also adults with PKU had insufficient intakes of PUFA (Moseley et al. 2002). In contrast, our study showed a sufficient intake of essential fatty acids, when all PUFA were taken together. Omega-3 supplements and fish-based bread spreads were widely used. Such products are common in Norwegian diets, and were not viewed as special to the PKU diet by the subjects or carers. More knowledge about intake and requirements for long-chain PUFA, in particular DHA status, is needed as this might influence cognitive outcome in PKU (Yi et al. 2011). In this study, however, only the dietary amounts of total PUFA seem relevant as subjects started treatment late.

The food recordings and serum amino acid profiles indicated that the subjects had a protein intake in accordance with requirements. This indicates that a low Phe diet supplemented with protein substitute will fulfil protein requirements for adult PKU patients. However, the amount of protein substitute necessary, coupled with the high fortification levels in these products, resulted in excessive intakes of most vitamins and trace minerals for these adults. This was reflected in high levels of folate in both erythrocytes and serum. Similar high folate levels were reported by Robinson et al. who assumed the high levels resulted from large amounts of vegetables in the PKU diet (Robinson et al. 2000). In our study, however, low vegetable consumption resulted in a low intake of natural folate. Even if the subjects had doubled their vegetable intake, the natural folate would be a minor fraction of the folic acid intake from protein substitutes. The intake and blood parameters reported indicate that the amounts of folic acid in three of the protein substitutes used are excessive for adult patients. The two subjects using Prekunil and Avonil had intakes of folic acid and vitamin B_{12} comparable to recommendations. They were the only subjects with erythrocyte and serum folate within normal range, see Table 4.

Natural sources of vitamin B_{12} are scarce in the PKU diet and several reports show that adults with PKU risk B_{12} deficiency if fortified protein substitutes are not taken (Hanley et al. 1993; Robinson et al. 2000; Hvas et al. 2006). In this study, all subjects took the substitutes as prescribed, and the high intake of vitamin B_{12} in most protein substitutes was reflected in the blood analyses, see Table 4. Vugteveen et al. report that vitamin B_{12} concentrations in serum within reference values, do not exclude functional vitamin B_{12} deficiency in PKU patients. In order to detect this, they recommend measuring serum methylmalonic acid or plasma homocysteine for these patients in the future (Vugteveen et al. 2011).

Despite a high iron intake, the blood analyses revealed no signs of iron overload. This indicates a low absorption of dietary iron in PKU diets. Similar lack of correlation is reported earlier in PKU children (Arnold et al. 2001; Acosta et al. 2004). Similar to reports by MacDonald, some subjects in our study complained of abdominal discomfort after taking the protein substitutes (MacDonald 2000). The possibility that excessive amounts of intestinal iron and other micronutrients contribute to these symptoms cannot be excluded.

Fortification of protein substitutes makes the diet less complicated and makes it easier to comply to and organise the diet, as shown by MacDonald (2000). The protein substitutes used in this study were recommended for older children, adolescents and adults. Our findings raise the question of whether a single product can meet the requirements for protein and micronutrients in children as well as adults. The amounts of micronutrients in the protein substitutes might need adjustment in order to meet vitamin and mineral recommendations in adult PKU patients. Further investigations, preferably in controlled studies with larger samples of adult PKU patients are necessary. Nutrient composition in the protein substitutes should be aimed at improving nutrient status in all adults adhering to a PKU diet.

In Summary

Despite methodological limitations in the present study, the data obtained showed that late-treated adult PKU patients manage to maintain a diet with adequate protein intake and therapeutic serum Phe levels over time. Subjects had, however, problems in adhering to nutritional recommendations for fruits and vegetables, added sugar and dietary fibre. Recommended intake of essential fatty acids required use of omega-3 supplements. Intake of folic acid, vitamin B_{12} and iron was very high, due to fortification of the protein substitutes. The food records were carefully and accurately done, showing excess intakes of micronutrients for all subjects using highly fortified protein substitutes. We presume that early and continuously treated adults with classical PKU may have similar intakes when low blood Phe levels are maintained by diet alone. Further studies are needed to determine if requirements for micronutrients and omega-3 fatty acids differ in PKU patients compared to the general population. In the meantime composition of protein substitutes intended for adults with PKU might need adjustment.

Acknowledgements We thank Susan Jane Sødal for assistance with the English language. We also send our gratitude to the patients and carers who participated in the study.

Conflict of Interest

Ingrid Wiig has attended meetings on dietetic management for metabolic disorders arranged and paid by Nutricia.

Kristina Motzfeldt has received compensation from Merck Serono as a member of the European Nutritionist Expert panel in PKU, and she has received fees for consulting or lecturing from Nutricia and Vitaflo Scandinavia.

References

Acosta PB, Yannicelli S, Singh RH, Elsas LJ, Mofidi S, Steiner RD (2004) Iron status of children with phenylketonuria undergoing nutrition therapy assessed by transferrin receptors. Genet Med 6:96–101

Arnold GL, Kirby R, Preston C, Blakely E (2001) Iron and protein sufficiency and red cell indices in phenylketonuria. J Am Coll Nutr 20:65–70

Baumeister AA, Baumeister AA (1998) Dietary treatment of destructive behavior associated with hyperphenylalaninemia. Clin Neuropharmacol 21:18–27

Fitzgerald B, Morgan J, Keene N, Rollinson R, Hodgson A, and Rymple-Smith J (2000) An investigation into diet treatment for adults with previously untreated phenylketonuria and severe intellectual disability. J Intellect Disabil Res 44: t-9

Hanley WB, Feigenbaum A, Clarke JT, Schoonheyt W, Austin V (1993) Vitamin B12 deficiency in adolescents and young adults with phenylketonuria. Lancet 342:997

Hvas AM, Nexo E, Nielsen JB (2006) Vitamin B12 and vitamin B6 supplementation is needed among adults with phenylketonuria (PKU). J Inherit Metab Dis 29:47–53

Lee PJ, Amos A, Robertson L, Fitzgerald B, Hoskin R, Lilburn M, Weetch E, Murphy G (2009) Adults with late diagnosed PKU and severe challenging behaviour: a randomised placebo-controlled trial of a phenylalanine-restricted diet 1. J Neurol Neurosurg Psychiatry 80:631–635

MacDonald A (2000) Diet and compliance in phenylketonuria. Eur J Pediatr 159(Suppl 2):136–141

MacDonald A, Rylance G, Davies P, Asplin D, Hall SK, Booth IW (2003) Free use of fruits and vegetables in phenylketonuria. J Inherit Metab Dis 26:327–338

McCaman MW, Robins E (1962) Fluorimetric method for the determination of phenylalanine in serum. J Lab Clin Med 59:885–890

Moseley K, Koch R, Moser AB (2002) Lipid status and long-chain polyunsaturated fatty acid concentrations in adults and adolescents with phenylketonuria on phenylalanine-restricted diet. J Inherit Metab Dis 25:56–64

Nordic Council of Ministers (2004) Nordic nutrition recommendations 2004. Nordic Council of Ministers, Aarhus

Poustie VJ, Wildgoose J (2010) Dietary interventions for phenylketonuria. Cochrane Database Syst Rev CD001304

Rimestad AH, Borgejordet Å, Vesterhus KN, Sygnestveit K, Løken EB, Trygg K, Pollestad ML, Lund-Larsen K, Omholdt-Jensen G, Nordbotten A (2001) Den store matvaretabellen (The official Norwegian table of food composition), 2nd edn. Gyldendal Norsk Forlag, Oslo

Robinson M, White FJ, Cleary MA, Wraith E, Lam WK, Walter JH (2000) Increased risk of vitamin B12 deficiency in patients with phenylketonuria on an unrestricted or relaxed diet. J Pediatr 136:545–547

Rose HJ, White F, MacDonald A, Rutherford PJ, Favre E (2005) Fat intakes of children with PKU on low phenylalanine diets. J Human Nutr Dietetics 18:395–400

Scriver CR, Levy H, Donlon D (2011) Hyperphenylalaninemia: phenylalanine hydroxylase deficiency. In: Valle D, Beaudet A, Vogelstein B, Kinzler KW, Antonarakis SE, Ballabio A (eds) The online metabolic & molecular bases of inherited disease. The McGraw-Hill Companies

Suitor CW, Bailey LB (2000) Dietary folate equivalents: interpretation and application. J Am Diet Assoc 100:88–94

Ten Hoedt AE, de Sonneville LM, Francois B, Ter Horst NM, Janssen MC, Rubio-Gozalbo ME, Wijburg FA, Hollak CE, Bosch AM (2011) High phenylalanine levels directly affect mood and sustained attention in adults with phenylketonuria: a randomised, double-blind, placebo-controlled, crossover trial. J Inherit Metab Dis 34:165–171

Trefz F, Maillot F, Motzfeldt K, Schwarz M (2011) Adult phenylketonuria outcome and management. Mol Genet Metab 104(Suppl):S26–S30

Vugteveen I, Hoeksma M, Monsen AL, Fokkema MR, Reijngoud DJ, van RM, van Spronsen FJ (2011) Serum vitamin B12 concentrations within reference values do not exclude functional vitamin

B12 deficiency in PKU patients of various ages. Mol Genet Metab 102:13–17

Weetch E, MacDonald A (2006) The determination of phenylalanine content of foods suitable for phenylketonuria. J Hum Nutr Diet 19:229–236

Yannicelli S, Ryan A (1995) Improvements in behaviour and physical manifestations in previously untreated adults with phenylketonuria using a phenylalanine-restricted diet: a national survey. J Inherit Metab Dis 18:131–134

Yi SH, Kable JA, Evatt ML, Singh RH (2011) A cross-sectional study of docosahexaenoic acid status and cognitive outcomes in females of reproductive age with phenylketonuria. J Inherit Metab Dis 34:455–463

Yi SH, Singh RH (2008) Protein substitute for children and adults with phenylketonuria. Cochrane Database Syst Rev CD004731

JIMD Reports
DOI 10.1007/8904_2012_158

RESEARCH REPORT

Did the Temporary Shortage in Supply of Imiglucerase Have Clinical Consequences? Retrospective Observational Study on 34 Italian Gaucher Type I Patients

Laura Deroma · Annalisa Sechi · Andrea Dardis ·
Daniela Macor · Giulia Liva ·
Giovanni Ciana · Bruno Bembi

Received: 09 May 2012 / Revised: 09 May 2012 / Accepted: 21 May 2012 / Published online: 1 July 2012
© SSIEM and Springer-Verlag Berlin Heidelberg 2012

Abstract *Background.* Enzyme Replacement Therapy (ERT) is the standard of care in Gaucher disease. The effects of withdrawal or reduced doses are debated, thus a retrospective cohort study was conducted to investigate clinical and laboratory differences in 34 Gaucher type 1 patients experiencing an ERT dosage reduction after the forced temporary imiglucerase shortage in 2009.

Methods. Haemoglobin concentration, leukocytes and platelets counts, and chitotriosidase activity were assessed at baseline and after 6 and 12 months (t0, t6, t12), while bone pain, energy, work or school performance, concentration, memory and social life every 3 months.

Results. The cohort was made up of 18 males and 16 females (medians: age 41.8 years, therapy duration 14.1 years, dosage reduction 35.5%). Haemoglobin, leukocytes and platelets remained substantially stable, while chitotriosidase activity showed an increase, especially after t6. Age, splenectomy or genotype were not associated with laboratory parameters changes, except for a significant median increase of chitotriosidase activity in non-splenectomised patients after 12 months ($p = 0.01$). At 3, 6, 9 and 12 months, more than 50% patients reported at least one problem in subjective well-being (56%, 65%, 70%, 58%, respectively), while bone pain occurred or worsened in 13/33, 13/32, 7/28 and 5/26 patients, respectively. No bone crises were reported.

Conclusions. Drug reduction did not induce substantial modification in the laboratory values but seems to have influenced the well-being perception of some Gaucher patients. Thus, bone pain, general health and quality of life should be carefully monitored during ERT reductions.

Background

Gaucher disease (GD), the most common lysosomal storage disorder, is a recessive autosomal disease due to mutations in the gene encoding the lysosomal enzyme acid beta glucosidase (*GBA1*). The deficient activity of this enzyme results in the accumulation of glucosylceramide (GlcCer) within the lysosomes. Type I (GD1) is the most common form of GD and may imply a large variety of symptoms, ranging from completely asymptomatic to child-onset forms (Beutler and Grabowski 2001; Jmoudiak and Futerman 2005).

Several therapeutic approaches are being examined with the aim to reduce the GlcCer intracellular burden, but Enzyme Replacement Therapy (ERT) with human recombinant acid beta glucosidase still remains the standard of care in these patients (Hughes and Pastores 2010).

In June 2009, the European Medicines Agency (EMEA) published a press release to communicate that after the viral contamination (calicivirus of the type Vesivirus 2117) of Genzyme's manufacturing plant in Allston Landing (USA), the company had to shut down the production of imiglucerase (Cerezyme ®, Genzyme Corporation, MA, USA) (EMEA 2009), causing the temporary worldwide shortage of the drug.

Communicated by: Verena Peters

Competing interests: None declared

L. Deroma (✉) · A. Sechi · A. Dardis · D. Macor · G. Liva ·
G. Ciana · B. Bembi
Regional Coordinator Centre for Rare Diseases, University Hospital
"Santa Maria della Misericordia", Piazzale Santa Maria della
Misericordia 15, Udine 33100, Italy
e-mail: deroma.laura@aoud.sanita.fvg.it; sechi.annalisa@aoud.sanita.fvg.it; dardis.andrea@aoud.sanita.fvg.it; macor.daniela@aoud.sanita.fvg.it; liva.giulia@aoud.sanita.fvg.it; ciana.giovanni@aoud.sanita.fvg.it; bembi.bruno@aoud.sanita.fvg.it

In 2009, imiglucerase was the only registered enzyme available for ERT in GD (Hollak et al. 2010); consequently, many patients under treatment were forced to withdraw or reduce doses.

So far, several studies have been published to describe the clinical consequences of this shortage (Giraldo et al. 2011; Goldblatt et al. 2011; Zimran et al. 2011). Moreover, although no trials were performed to ascertain the possible effects of a temporary dosage reduction or withdrawal, this issue had already been dealt with before the occurrence of imiglucerase shortage (Drelichman et al. 2007; Elstein et al. 2000; Grinzaid et al. 2002; Schwartz et al. 2001; Vom Dahl et al. 2001). However, each study analysed only a few patients and the results among the studies were not homogeneous.

The aim of this study was to assess possible differences in selected laboratory values and clinical aspects in patients experiencing an ERT dosage reduction after the temporary imiglucerase shortage in 2009.

Methods

A retrospective cohort study was conducted on a group of GD1 patients. The inclusion criteria were having experienced a drug reduction due to imiglucerase shortage in 2009 for at least 1 year and being followed up at the Regional Coordinator Centre for Rare Diseases of the University Hospital "Santa Maria della Misericordia" (Udine, Italy) until July 2011.

The haemoglobin concentration (Hb) and white blood cells (WBC) and platelets (Plt) counts were examined. Chitotriosidase (Ct) activity, acknowledged as a marker of disease (Hollak et al. 1994), was measured as previously described (Hollak et al. 1994). These parameters were evaluated no more than 6 months before the drug reduction and after 6 and 12 months (t0, t6, t12, respectively). Since a closer observation was performed after the shortage (a phone call was made by the charge nurse every 3 months for 1 year), it was also possible to evaluate their subjective condition after the drug dosage reduction. In particular, patients were asked about the occurrence or the worsening of bone pain. Several specific questions (five for the adults and three for the paediatric patients) were also asked about a decline in vitality/strength, work/school performance, concentration/memory and social life after the dosage decrease.

The Ethics Committee of the University Hospital "Santa Maria della Misericordia" (Udine, Italy) approved this study.

Statistical Analysis

Continuous variables are described as median and first and third quartile. Categorical variables are presented as frequency and percentage. The Shapiro Wilk test was used to check the normality assumption and since the variables were not normally distributed, the signed rank test was used to compare two groups. All the analyses were performed using the statistical package Stata 11.0 (Stata Statistical Software: Release 11.0, 2009. StataCorp LP, College Station, TX, USA).

Results

Thirty-four out of the 38 Gaucher patients followed up until July 2011 fulfilled the inclusion criteria and constituted the study retrospective cohort. Eighteen patients were males and 16 females, the median age was 41.8 years and 4 of them were children (age < 16 years). Their main characteristics are described in Table 1.

All patients were treated with imiglucerase when the drug shortage occurred and all experienced one or two drug reductions in July–August 2009. Their pre-reduction median dosage was 55.5 units/kg/month, while the median dosage after the decrease was 15.8 units/kg/month (signed rank test $p < 0.0001$). The median percentage variation between the pre- and post-reduction doses was 35.5% (Table 1).

Laboratory Values

Levels of Hb, WBC, Plt and Ct activity at baseline and at 6 and 12 months after drug reduction are summarised in

Table 1 Characteristics of the patients

Variables	n (%)	Median (IQR)
Gender		
Male	18 (53)	
Female	16 (47)	
Genotype		
N370S/N370S	5 (15)	
N370S/other	18 (53)	
Other/other	11 (32)	
Previous spleen removal		
Yes	10 (29)	
No	24 (71)	
Age, at therapy start (years)		27.8 (13.2–37.3)
Age, at reduction (years)		41.8 (29.2–50.5)
Duration of therapy, at reduction (years)		14.1 (10.5–15.6)
ERT dose (IU/kg/month)		
Before reduction		55.5 (48–63)
After reduction		15.8 (15–30)
% variation		35.5 (28–47)

Table 2. Results of the comparisons t0-t6, t6-12, t0-t12 (signed rank test) are also reported.

Hb concentration did not show important variations between t0 and t6 (median variation 0.1 g/dL, IQR $-0.3 \div 0.5$, $n = 22$), t6 and t12 (median -0.4 g/dL, IQR $-0.7 \div 0.1$, $n = 16$) and t0-t12 (median -0.1, IQR $-0.7 \div 0.4$, $n = 23$). A decrease was evidenced in 7/22 patients at t6 and in 14/23 at t12, but none ever suffered from anaemia, as defined by Pastores et al. (2004).

Overall, the WBC and platelets counts after 6 and 12 months were not statistically different from the baseline values (Table 2). When compared to t0, WBC showed a decrease in 8/22 patients at t6 and in 12/24 patients at t12, while platelets were reduced at t6 in 10/22 patients and at t12 in 13/22. However, only one reached a value lower than 100,000/µL (89,000 at t6).

When only the patients without missing values in the measurements of Hb ($n = 16$), WBC ($n = 17$) and Plt ($n = 16$) were analysed separately, the results did not change.

Chitotriosidase activity significantly increased between t6 and t12 ($p = 0.0004$), with a median variation of 398 nmol/mL/h (IQR $55 \div 590$) in the 24 patients examined, but did not vary between t0 and t6 (median 138, IQR $-383 \div 280$, $n = 20$) nor between t0 and t12 (median 311, IQR $-49 \div 634$, $n = 21$). Nevertheless, when the patients without missing values were analysed ($n = 20$), Ct after 6 months (median 703; IQR 511–802) was not different from Ct at t0 (median 555; IQR 334–940), but statistically significant increases were noticed between t0 and t12 (median 1,114; IQR 516–1,391; $p = 0.04$) and between t6 and t12 ($p = 0.01$).

Comparisons were also made to assess whether being a child, being splenectomised or having a particular genotype (N370S/N370S vs N370S/other and other/other; N370S/N370S and N370S/other vs other/other) might affect the response to ERT shortage. No differences were found,

except when Ct activity at t0-t12 was compared between splenectomised and non-splenectomised patients. While splenectomised patients showed a non-significant median decrease from 1,322 to 834 nmol/mL/h ($p = 0.17$), the non-splenectomised patients showed a significant median increase from 543 to 1,191 nmol/mL/h ($p = 0.01$).

Subjective Well-Being

Table 3 shows the number of patients who reported a worsening in selected aspects of their life (energy, work or school performance, concentration, memory, social life) at different times after the drug reduction (3, 6, 9, 12 months). More than 50% of patients declared at least one subjective problem that arose 3, 6, 9 and 12 months after the drug reduction (56%, 65%, 70%, 58%, respectively).

Figure 1 shows the distribution of the number of problems that occurred after the drug reduction and reported at the four different time points by the adult patients who answered the five questions on subjective

Table 3 Problems that impaired the perception of well-being after drug reduction, as reported by patients (compared to t0)

Problems	t3	t6	t9	t12
↓Energy	15/33	12/32	11/28	3/26
↓Performance (work/school)	7/33	7/32	3/28	4/26
↓Concentration[a]	4/29	8/29	9/24	8/24
↓Memory[a]	6/29	14/29	11/24	9/24
↓Social life[a]	4/28	5/28	3/25	2/24
↑Tiredness[b]	1/4	1/3	1/3	0/2
≥ 1 problem	19/33	21/32	20/28	15/26

[a] Only adults
[b] Only children

Table 2 Laboratory values at t0, t6 and t12 and their differences between t0-t6, t6-12, t0-t12

Parameter (unit)	t0 Median (IQR) n	t6 Median (IQR) n	t12 Median (IQR) n	t0-t6 p-value n	t6-t12 p-value n	t0-t12 p-value n
Haemoglobin (g/dL)	14.0 (13.1–15.2) $n = 30$	14.6 (12.9–15.5) $n = 22$	14.1 (13.1–15.0) $n = 25$	0.28 $n = 22$	0.02 $n = 16$	0.20 $n = 23$
White blood cells (n/µL)	5550 (4690–7255) $n = 30$	5460 (4780–7940) $n = 22$	6130 (5080–7790) $n = 26$	0.29 $n = 22$	0.96 $n = 17$	0.45 $n = 24$
Platelets (10^3/µL)	193 (150–292) $n = 30$	198 (149–301) $n = 22$	231 (152–292) $n = 24$	0.75 $n = 22$	0.78 $n = 16$	0.97 $n = 22$
Chitotriosidase (nmol/mL/h)	568 (349–919) $n = 21$	642 (467–787) $n = 28$	1,057 (543–1405) $n = 25$	0.88 $n = 20$	0.0004 $n = 24$	0.19 $n = 21$
Chitotriosidase (nmol/mL/h) (patients without missing values)	555 (334–940) $n = 20$	703 (511–802) $n = 20$	1,114 (516–1391) $n = 20$	0.88 $n = 20$	0.04 $n = 20$	0.01 $n = 20$

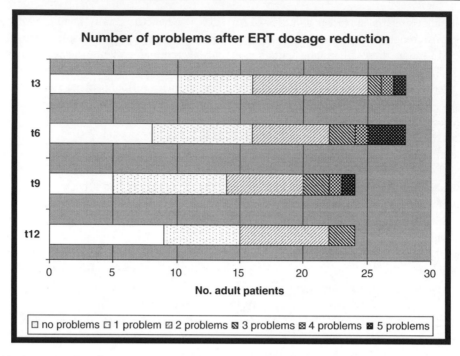

Fig. 1 Number of subjective problems reported by adult patients after the drug reduction. *Note: only patients who answered all five questions*

well-being. As for children, at t3, one child out of four reported one problem (tiredness), at t6 and t9, one child out of three reported two problems (tiredness and loss of energy in the same child), while at t12 none of the two children reported problems.

Bone Pain

Thirty-three patients answered the questions about bone pain; among these, 13 (2 children) already suffered from bone pain before the drug reduction. The frequency of bone pain onset in asymptomatic patients and its worsening in patients already suffering from bone symptoms is reported in Table 4. No patients suffered from bone crises.

Discussion

The aim of this study was to investigate whether the imiglucerase dosage reduction due to the temporary shortage in supplies had an effect on the clinical conditions and laboratory parameters of 34 GD1 patients, four of whom were children.

This is the only Italian study investigating the possible consequences of this forced ERT dose reduction. Furthermore, even when considering the international literature, only one previous study (Giraldo et al. 2011), focusing on the effects of a dosage reduction rather than those of a complete withdrawal, was performed. These authors described a group of 17 Spanish GD patients and reported

Table 4 Onset and worsening of bone pain after drug reduction (compared to t0)

Bone pain	t3	t6	t9	t12
Onset				
Overall	5/20	6/19	4/18	4/17
Adults	5/18	6/17	4/16	4/15
Children	0/2	0/2	0/2	0/2
Worsening				
Overall	8/13	7/13	3/10	1/9
Adults	8/11	6/12	2/9	1/9
Children	0/2	1/1	1/1	0/0
Onset or worsening				
Overall	13/33	13/32	7/28	5/26
Adults	13/29	12/29	6/25	5/24
Children	0/4	1/3	1/3	0/2

stable values of Hb and Plt 6 months after ERT reduction, consistently with the results of the present study. However, a statistically significant increase of the chitotriosidase activity was reported ($p = 0.03$) after 6 months (Giraldo et al. 2011), while the increase observed in the Italian patients at the same time point did not reach statistical significance but became significant between 6 and 12 months. The increase was even more evident when only patients without missing values were taken into account, since Ct increased from 555 at baseline to 703 at 6 months and 1,114 at 12 months.

About 40 % (7/17) of the Spanish patients complained of diffuse bone pain (Giraldo et al. 2011), consistent with the fraction of patients of the present study that reported the onset or worsening of bone pain at 3 and 6 months. While no bone crises were observed in the Italian patients, Giraldo et al. reported a bone crisis in three cases (Giraldo et al. 2011). This difference could be due to a different disease severity in the two groups of patients, the Italian cohort being milder.

As for ERT withdrawal, Elstein et al. reported stable clinical conditions and laboratory values in 15 patients after 4 years of follow-up (Elstein et al. 2000), consistent with the results of a Brazilian case report describing clinical stability after a 3-month withdrawal (Schwartz et al. 2001). However, several authors reported the deterioration of haematological parameters (Hb, Plt, Ct) and an increase in organomegaly (Zimran et al. 2011; Grinzaid et al. 2002; Vom Dahl et al. 2001) after an ERT withdrawal lasting from 3 to more than 24 months. Moreover, Giraldo et al. reported conflicting results on 23 patients who withdrew ERT for 6 months, describing stable Hb and Plt in the great number of patients but at the same time the occurrence of a bone crisis in one patient and of mild anaemia in another one (Giraldo et al. 2011). Two recent studies (Giraldo et al. 2011; Goldblatt 2011) showed that even if laboratory parameters remained stable in most GD patients after 5–6 month of ERT withdrawal, some of them worsened their clinical condition.

Therefore, it may be argued that some characteristics of GD patients could play a role in the response to dosage variation. Thus, in this study, the possible role of genotype, age and splenectomy on the laboratory parameters after ERT reduction was investigated. Among the four patients described by Grinzaid et al., a better clinical and laboratory behaviour was seen in the one with a N370S/N370S genotype after a therapy discontinuation of ≥ 1 year (Grinzaid 2002). On the contrary, among the Italian patients, no differences were found in laboratory values variations according to genotype.

No differences were even found when comparing paediatric and adult patients, in contrast with a study conducted in Argentina that reported a clear deterioration of the clinical condition in five children forced to withdraw for 15–36 months (Drelichman et al. 2007).

Finally, the non-splenectomised patients showed a significant increase in chitotriosidase activity between t0 and t12, while in splenectomised patients a non-significant decrease was observed. This is in contrast with the study by Czartoryska et al., reporting a more evident increase in chitotriosidase activity after ERT cessation in two splenectomised patients when compared to six non-splenectomised patients (Czartoriska et al. 2000). However, the reduced Ct activity in splenectomised patients could be explained with the spleen removal itself, which leads to a reduction of the number of macrophages and, in turn, to their Ct production.

In spite of the substantial maintenance of laboratory values, more than one half of the studied patients reported at least one problem that worsened their perception of well-being during the follow-up period. Moreover, a variable number of patients reported the onset or worsening of bone pain (from about 40 % at t3 and t6 to about 20% at t12). However, consideration of the psychological impact of experiencing a forced reduction of a drug, that is thought to be indispensable by the patient, cannot be underestimated. In fact, the majority of these patients had good control of GD under ERT treatment for years, being allowed to live normal lives. When the shortage occurred, many patients reported to be worried about a possible influence on the outcome of their disease. Therefore, the subjective problems they reported could be influenced by this psychological attitude. Finally, it is important to point out that the information on both well-being and bone pain was not gathered using a validated questionnaire and it is difficult to evaluate the relevance of the problems reported.

Conclusions

In summary, the ERT dosage reduction did not cause important changes in the laboratory parameters, except for an increase of the Ct activity. Nevertheless, a part of GD patients experienced some modification in their general conditions. Therefore, when a decrease in ERT dosage needs to be introduced, a careful evaluation of the general health (with a special attention to bone pain) and quality of life of each patient should be carried out.

Synopsis

Bone pain, general health and quality of life should be carefully monitored during ERT reductions.

Authors' Contributions

LD participated in the study design, performed the statistical analysis and drafted the manuscript. AS participated in the study design and in the draft of the manuscript. AD carried out the chitotriosidase activity and genotype analyses and participated in the draft of the manuscript. DM acquired the data and participated in the draft of the manuscript. GL acquired the data and participated in the draft of the manuscript. GC participated in the design of the study. BB conceived the study, participated in its design and coordination and in the draft of the manuscript. All authors read and approved the final manuscript.

Competing Interests

The authors declare that they have no competing interests.

References

Beutler E, Grabowski GA (2001) Gaucher disease. In: Scriver CR, Beaudet AL, Valle D, Sly WS, Childs B, Kinzler KW, Vogelstein B (eds) The metabolic & molecular bases of inherited disease. McGraw-Hill Medical Publishing Division, New York, pp 3635–3668

Czartoriska B, Tylki-Szymanska A, Lugowska A (2000) Changes in serum chitotriosidase activity with cessation of replacement enzyme (cerebrosidase) administration in Gaucher disease. Clin Biochem 33(2):147–149

Drelichman G, Ponce E, Basack N et al (2007) Clinical consequences of interrupting enzyme replacement therapy in children with type 1 Gaucher disease. J Pediatr 151:197–201

Elstein D, Abrahamov A, Hadas-Halpern I, Zimran A (2000) Withdrawal of enzyme replacement therapy in Gaucher's disease. Br J Haematol 110:488–492

European Medicines Agency (EMEA) (2009) Supply shortages of Cerezyme and Fabrazyme – priority access for patients most in need of treatment recommended. Doc. Ref. EMEA/389995/2009. London, 25-6-2009

Giraldo P, Irun P, Alfonso P et al (2011) Evaluation of Spanish Gaucher disease patients after a 6-month imiglucerase shortage. Blood Cells Mol Dis 46:115–118

Goldblatt J, Fletcher JM, McGill J, Szer J, Wilson M (2011) Enzyme replacement therapy "drug holiday": results from an unexpected shortage of an orphan drug supply in Australia. Blood Cells Mol Dis 46:107–110

Grinzaid KA, Geller E, Hanna SL, Elsas LJ II (2002) Cessation of enzyme replacement therapy in Gaucher disease. Genet Med 4 (6):427–433

Hollak CEM, van Weely S, van Oers MHJ, Aerts JMFG (1994) Marked elevation of plasma chitotriosidase activity. A novel hallmark of Gaucher disease. J Clin Invest 93:1288–1292

Hollak CEM, vom Dahl S, Aerts JMFG et al (2010) Forze majeure: therapeutic measures in response to restricted supply of imiglucerase (Cerezyme) for patients with Gaucher disease. Blood Cells Mol Dis 44:41–47

Hughes DA, Pastores GM (2010) The pathophysiology of GD – current understanding and rationale for existing end emerging therapeutic approaches. Wien Med Wochenschr 160:594–599

Jmoudiak M, Futerman AH (2005) Gaucher disease: pathological mechanisms and modern management. Br J Haematol 129:178–188

Pastores GM, Weinreb NJ, Aerts H et al (2004) Therapeutic goals in the treatment of Gaucher disease. Semin Hematol 41(Suppl 5):4–14

Schwartz IVD, Karam S, Ashton-Prolla P et al (2001) Effects of imilglucerase withdrawal on an adult with Gaucher disease. Br J Haematol 113:1089

Vom Dahl S, Poll LW, Haussinger D (2001) Clinical monitoring after cessation of enzyme replacement therapy in M Gaucher. Br J Haematol 113:1084–1085

Zimran A, Altarescu G, Elstein D (2011) Nonprecipitous changes upon withdrawal from imiglucerase for Gaucher disease because of a shortage in supply. Blood Cells Mol Dis 46:111–114

JIMD Reports
DOI 10.1007/8904_2012_166

5-Oxoprolinuria in Heterozygous Patients for 5-Oxoprolinase (*OPLAH*) Missense Changes

Eduardo Calpena · Mercedes Casado ·
Dolores Martínez-Rubio · Andrés Nascimento ·
Jaume Colomer · Eva Gargallo ·
Angels García-Cazorla · Francesc Palau ·
Rafael Artuch · Carmen Espinós

Received: 24 April 2012 / Revised: 04 June 2012 / Accepted: 14 June 2012 / Published online: 6 July 2012
© SSIEM and Springer-Verlag Berlin Heidelberg 2012

Abstract The inherited 5-oxoprolinuria is primarily suggestive of genetic defects in two enzymes belonging to the gamma-glutamyl cycle in the glutathione (GSH) metabolism: the glutathione synthetase (GSS) and the 5-oxoprolinase (OPLAH). The GSS deficiency is the best characterized of the inborn errors of GSH metabolism, whereas the OPLAH deficiency is questioned whether it is a disorder or just a biochemical condition with no adverse clinical effects. Recently, the first human *OPLAH* mutation (p.H870Pfs) was reported in homozygosis in two siblings who suffered from 5-oxoprolinuria with a benign clinical course. We report two unrelated patients who manifested massive excretion of 5-oxoproline in urine. In both probands, the blood GSH levels were normal and no mutations were found in the *GSS* gene. The mutational screening of the *OPLAH* gene, which included the codified sequences, the intronic flanking sequences, the promoter sequence, and a genetic analysis in order to detect large deletions and/or duplications, showed that each patient only harbors one missense mutation in heterozygosis. The in silico analyses revealed that each one of these *OPLAH* mutations, p.S323R and p.V1089I, could alter the proper function of this homodimeric enzyme. In addition, clinical symptoms manifest in these two probands were not related to GSH cycle defects and, therefore, this study provides further evidence that oxoprolinuria may present as epiphenomenon in several pathological conditions and confound the final diagnosis.

Communicated by: Verena Peters

Competing interests: none declared

E. Calpena · D. Martínez-Rubio · F. Palau
Genetics and Molecular Medicine Unit, Instituto de Biomedicina de Valencia – CSIC and CIBER de Enfermedades Raras (CIBERER), Valencia, Spain

M. Casado · A. Nascimento · J. Colomer · E. Gargallo ·
A. García-Cazorla · R. Artuch
Clinical Biochemistry, Pediatric Neurology and Pediatrics Departments, Hospital Sant Joan de Déu and CIBER de Enfermedades Raras (CIBERER), Barcelona, Spain

C. Espinós
Neurogenetics Platform. CIBER de Enfermedades Raras (CIBERER) and Instituto de Investigación Sanitaria La Fe., Valencia, Spain

C. Espinós (✉)
Hospital U. i P. La Fe, Escuela de Enfermería. Sótano., Avd de Campanar, 21, 46009, Valencia, Spain
e-mail: cespinos@ciberer.es

Introduction

There are several metabolic disturbances that are questioned if they only are a casual finding during biochemical screening of patients or have a direct clinical involvement. One of them is the 5-oxoprolinuria (pyroglutamic aciduria) whose etiology remains elusive in some circumstances. Several diseases and environmental conditions (special diets, drug metabolism and drug treatment, prematurity, malnutrition, some inborn errors of metabolism) have been associated with oxoprolinuria (Schwahn et al. 2005; Ruijter et al. 2006; Ristoff and Larsson 2007), although the hereditary 5-oxoprolinuria is primarily suggestive of genetic defects in two enzymes belonging to the gamma-glutamyl cycle in the glutathione (GSH) metabolism: the glutathione synthetase (GSS; EC 6.3.2.3) and the 5-oxoprolinase (OPLAH; EC 3.5.2.9).

The glutathione synthetase deficiency (MIM 266130) is the best characterized and the most common of the inborn errors of GSH metabolism (Njalsson 2005; Ristoff and Larsson 2007). It presents with a wide spectrum of clinical signs. The most frequent hallmarks are 5-oxoprolinuria, metabolic acidosis and hemolytic anemia, and in severely affected patients, the central nervous system is affected. This deficiency is caused by mutations in the *GSS* gene and more than 30 mutations have been described which are transmitted in an autosomal recessive fashion (Al-Jishi et al. 1999; Dahl et al. 1997; Njalsson et al. 2003; Shi et al. 1996). Heterozygous carriers of *GSS* mutations are healthy and show an enzyme activity of 55% of the normal mean and normal levels of GSH (Njalsson et al. 2005). However, the 5-oxoprolinase deficiency (MIM 260005) is questioned whether it is a disorder or just a biochemical condition with no adverse clinical effects. To date, nine probands with 5-oxoprolinuria and a low activity of 5-oxoprolinase have been described worldwide (Almaghlouth et al. 2011; Bernier et al. 1996; Cohen et al. 1997; Henderson et al. 1993; Larsson et al. 1981; Mayatepek et al. 1995; Roesel et al. 1981). These patients lack a consistent clinical picture except for the 5-oxoprolinuria. Symptoms reported in individual patients include renal stone formation, mental retardation, neonatal hypoglycaemia, microcytic anemia, and microcephaly. The 5-oxoprolinase is encoded by the *OPLAH* gene, and mutations in this gene are expected to lead to a 5-oxoprolinase deficiency transmitted in an autosomal recessive manner. Recently, the first human *OPLAH* mutation (p.H870Pfs) was reported, which predicts a truncated protein. This change was identified in homozygosis in two siblings with a persistent increased 5-oxoproline excretion and with a benign clinical course (Almaghlouth et al. 2011).

Here we report two new unrelated patients who manifested a massive excretion of 5-oxoproline in urine. After an exhaustive genetic analysis, in each patient, only one heterozygous missense mutation was identified in the *OPLAH* gene, suggesting that only one mutation could alter the normal activity of this homodimeric enzyme. Moreover, clinical features in these two probands were not related supporting the 5-oxoprolinase deficiency is a benign biochemical condition.

Materials and Methods

Patients This study protocol was approved by the ethics committee of the Hospital Sant Joan de Déu. Written informed consent was obtained from parents of the patients.

Case 1 The patient is a 1.5-month-old girl (ID no. 943, family AR-141; Fig. 1a), the first child of healthy non-consanguineous parents from Indian origin, with an unevent-

ful pregnancy and delivery. She was admitted in our hospital due to repeated episodes of choking (respiratory difficulties and perioral cyanosis shortly after breast feeding) and generalized tonic-clonic seizures. First baseline tests revealed hypocalcemia (1.4 mmol/L; control values: 2.1–2.7), increased phosphate values (2.8 mmol/L: control values 1.4–2.2), 25 hydroxyvitamin D deficiency (5.5 ng/mL: control values 12–62), increased PTH values (15.6 pmol/L: control values 0.5–5.5) normocytic normochromic anemia (hemoglobin: 8.5 g/dL: control values 9.0–13.0) with low plasma cobalamin (139 pmol/L: control values >198), normal plasma total homocysteine values (3.8 μmol/L: control values < 7.5), and metabolic acidosis. EEG and brain ultrasound were normal. General and neurological examinations were normal between episodes. The patient was treated with oral vitamin D supplementation and intramuscular vitamin B12 due to a suspicion of deficiency related to maternal vegetarian diet. The outcome was excellent with no other seizure recurrence, normalization of all altered biochemical parameters, and normal developmental delay at 1 year of age. Urine organic acids were analyzed three times, disclosing increased pyroglutamic acid excretion twice during decompensation (7,828 mmol/mol creatinine and 3,255 mmol/mol creatinine) which led us to the suspicion of pyroglutamic aciduria. These values were completely normalized at 1 year of age (9 mmol/mol creatinine; reference values <10). No causes of secondary increased excretion of pyroglutamic acid were detected including special diets, drugs (vigabatrin paracetamol and antibiotics), inborn errors of metabolism (urea cycle defects, cystinosis, tyrosinemia, and homocystinuria) and low glycine values secondary to malnutrition. Blood glutathione levels were normal (3.2 mmol/L; reference values: 1.5–3.1).

Case 2 An 8-year-old boy (ID no. 1037, family AR-152; Fig. 1a) diagnosed of Duchenne muscular dystrophy (DMD; MIM 310200) harboring a deletion that encompasses exons 46 to 51 at the *dystrophin* gene. Familiar antecedents were uneventful. Clinical presentation at the moment of exploration was classical, with muscular involvement and mental disability. No other symptoms or signs were present. Blood count (no anemia) and other routine laboratory parameters were normal besides increased serum creatine kinase activity. Urine organic acids were collected to study metabolomic profile by gas chromatography mass spectrometry, and we encountered a serendipitous finding of increased excretion of pyroglutamic acid (13,208 mmol/mol creatinine: reference values: <10). This huge excretion was further confirmed in a second urine sample (7,931 mmol/mol creatinine). In that moment, patient was under corticoid therapy (prednisone, 20 mg/day). No causes of secondary increased excretion of pyroglutamic acid were detected. Blood GSH values in were normal (2.45 mmol/L, reference values: 1.5–3.1).

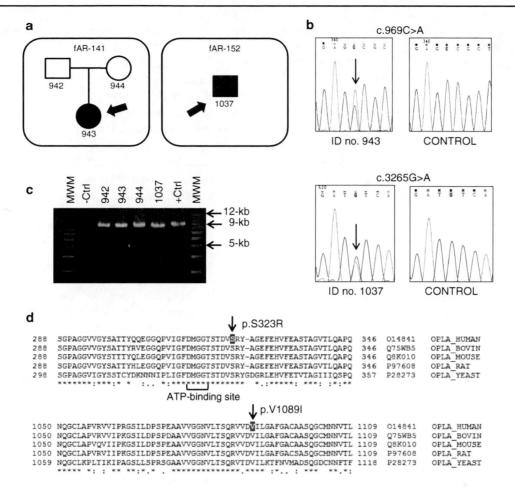

Fig. 1 Genetic findings. (**a**) Available DNAs from probands and their relatives. (**b**) Electrophoregrams show the identified mutations in proband ID no. 943 (c. 969C>A, p.S323R) and in proband ID no. 1037 (c.3265G>A, p.V1089I) in the *OPLAH* gene, together with the corresponding controls' electrophoregrams. (**c**) Electrophoresis gel shows the obtained band of 9,0-kb which corresponds to the amplification of the *OPLAH* gene. No differences were found in any of the analyzed samples ruling out large deletions or duplications (1st and 8th lanes: MWM, Molecular Weight Marker, 1-kb plus DNA ladder; 2nd lane: –Ctrl, PCR negative control; 3rd to 6th lanes: analyzed DNAs, 942, 943, and 944 from family AR-141, and 1037 from family AR-152; 7th lane: +Ctrl, DNA from a healthy subject). (**d**) Alignment of sequences of the OPLAH protein from human, bovin, mouse, rat, and yeast showing that both mutated residues are conserved amino acids. Identical amino acids in all of these sequences are indicated with an asterisk (*). The closest ATP-binding site to the p.S323R change is shown, which consists of the sequence DMGGT (from residue 324 to 328 in yeast)

Biochemical analyses Organic acids in urine were analyzed by gas chromatography/mass spectrometry (GC-MS) (Agilent Technologies Inc., Santa Clara, CA, USA) after extraction of the urine sample with ethylacetate and diethyleter and derivatization with *N,N*-bis(trimethylsilyl) trifluoroacetamide. The concentration of pyroglutamic acid was quantified by comparing the signal of m/z 156 obtained with standard solutions of the pure compound, using undecanoate as internal standard.

Genetic analyses A search for mutations was performed by Sanger sequencing of the PCR products of exons and their intronic flanking sequences in the *GSS* gene (NM_000178) as well as in the *OPLAH* gene (NM_017570.3) in an ABI Prism 3130xl autoanalyzer (Applied Biosystems, Foster City, CA, USA). The promoter sequence was also analyzed in the *OPLAH* gene.

With the aim to discard the possible existence of large deletions or insertions, we amplified all the codified exons and introns of the *OPLAH* gene (9,007 bp) in the two probands, the available relatives (ID no. 943's parents), and three control subjects. The PCR was performed using Long PCR Enzyme Mix (Mbi Fermentas, Glen. Burnie, MD, USA) and the following primers: forward 5'-GTGGGTCTCTCCCTCAGGAACC-3', and reverse 5'-CTGCAGCTCCGAGTCTCAGTGTC-3'. The amplified product was resolved by 0.8 % agarose electrophoresis. To improve the resolution of this study, the PCR products were separately digested with two enzymes, *Bam*HI and *Xho*I, which cut four or three times respectively. The digested fragments were under 5,500 bp and they were resolved by 0.8 % agarose electrophoresis.

Each identified change was also analyzed in more than 250 chromosomes from healthy individuals using DHPLC (Denaturing High Performance Liquid Chromatography; Transgenomic WAVE, Crewe, UK). We also investigated in silico the biological relevance of the mutated residues. Conservation of residues was analyzed by alignment of related sequences using the program BLAST (http://blast.ncbi.nlm.nih.gov/Blast.cgi) (Altschul et al. 1990). Sequence-based predictions of the phenotypic consequences of mutation were assessed using the SIFT (http://sift.jcvi.org/) and the PolyPhen (http://genetics.bwh.harvard.edu/pph/) softwares.

Results

No mutations were detected in the *GSS* gene in any of the two probands. In patient ID no. 943, a novel change was detected in the *OPLAH* gene: c.969C>A (Fig. 1b), which is predicted to generate the novel p.S323R amino acid change. We also analyzed her parents, and the mother was a heterozygous carrier of the p.S323R change whereas the father did not harbor it. In the three available samples, we discarded the existence of large deletions or duplications in the *OPLAH* gene (Fig. 1c). The p.S323R was not detected in 128 control subjects. The residue S323 is an evolutionary conserved amino acid, invariant across more than 100 different species (Fig. 1D). Both SIFT and PolyPhen algorithms predicted that the p.S323R change would be probably damaging.

Patient ID no. 1037 was carrier of the *OPLAH* c.3265G>A change (Fig. 1b), which is predicted to produce the p.V1089I amino acid change. Large deletions and/or duplications of the *OPLAH* gene were ruled out in this patient (Fig. 1c). A search for the p.V1089I change was performed in 154 healthy individuals and this was observed as a heterozygous change in three individuals. In fact, this change is annotated as SNP rs185836803 in dbSNP database (http://www.ncbi.nlm.nih.gov/SNP/snp_ref.cgi?rs=185836803), although information about heterozygosity is not available. The residue V1089 is a relatively conserved amino acid (Fig. 1d). Results obtained from the in silico analysis were controversial: the SIFT software predicted that p.V1089I could be tolerated and the PolyPhen algorithm that this would be damaging.

Discussion

Several inborn errors of GSH metabolism are known, glutathione synthetase deficiency being the first cause of GSH cycle diseases (Shi et al. 1996). GSH takes part in multiple fundamental processes apart of being one of the most important antioxidants in the eukaryotic organism.

Deficiency of glutathione synthetase could affect the central nervous system and other tissues and may cause neurological involvement and anemia, among other signs. Thus, the massive excretion of 5-oxoproline in our patients could be attributed to mutations in the *GSS* gene, although GSH values in blood were within normal limits and no mutations were identified in the *GSS* gene in any of our two probands, ruling out this diagnosis.

The deficiency of 5-oxoprolinase has been associated with a wide spectrum of clinical symptoms (Ristoff and Larsson 2007). To date only one homozygous mutation has been described in two siblings who show a benign clinical course but with a persistent increased excretion of 5-oxoproline (Almaghlouth et al. 2011). The proband ID no. 943 presented with an increased urinary excretion of 5-oxoproline at the beginning of the clinical picture, which spontaneously normalized at 1 year of age, together with a complete recovery of her clinical picture. In this case, we cannot rule out that low vitamin D and cobalamin levels could reflect a malnutrition status that could contribute to the pyroglutamic aciduria. Nevertheless, the presence of the reported mutation was probably necessary for triggering the pyroglutamic aciduria, since no hyperhomocysteinemia was revealed in this patient and plasma glycine values were always normal. The proband ID no. 1037 was a patient with a classical Duchenne clinical presentation, with no other signs that could be attributed to oxoprolinuria. In each of them, only one missense change in heterozygosis was identified in the *OPLAH* gene after analyzing the codified exons, their intronic flanking regions, and the promoter sequence, and discarding large deletions/insertions. Thus, another mutation in the *OPLAH* gene is improbable except located in a deep intronic region. Splicing mutations account for around 10 % of all reported mutations and deep intronic mutations represent less than 1 % of known splicing mutations (Cooper et al. 2010). All these data together suggest that these two probands seem to be carriers of only one defect in the *OPLAH* gene.

Both mutations were detected in healthy subjects: p.V1089I in three controls and the p.S323R in the proband ID no. 943's mother. The clinical course of the two patients with known mutations in the *OPLAH* gene is largely benign (Almaghlouth et al. 2011), as confirmed in our first case. Concerning the second case, DMD is an X-linked neuromuscular disorder due to mutations in the gene encoding dystrophin, and therefore has a well-defined etiology. Patients diagnosed of DMD and also with oxoprolinuria have not been reported. We tested pyroglutamic acid excretion in urine in five further patients with DMD, but disclosed normal results (data not shown). Therefore, persistent oxoprolinuria in the proband ID no. 1037 does not seem to be related to the DMD. Two female siblings were reported with a severe neurological picture, but only one of them manifested oxoprolinuria (Cohen et al. 1997). In that case, authors concluded that both siblings suffered from an unknown hereditary disease, unrelated to the 5-oxoprolinase deficiency observed in one of them. In the aggregate, our study supports the proposition that deficiency of 5-oxoprolinase is a benign condition. For this reason, mutations in the *OPLAH* gene could remain undiscovered since they probably would not lead to relevant clinical symptoms but may cause biochemical alterations that confound the final diagnosis.

This study reveals that heterozygous missense mutations in the *OPLAH* gene could cause oxoprolinuria. The in silico analyses show that both mutations could be pathological. For both changes, the sequence alignment revealed that the affected residues are conserved throughout evolution, being identical in mammals and yeast (Fig. 1d). At least one of the algorithms used to predict the possible pathogenicity revealed that each one of these mutations could be damaging. Little is known about the functionality of the human 5-oxoprolinase. In *Saccharomyces cerevisiae*, the *OXP1/YKL215c* encodes for an ATP-dependent 5-oxoprolinase (Kumar and Bachhawat 2010). This enzyme functions as a dimer and contains two distinct domains. The first domain, from residue 1 to 736, is the actin-like ATPase motif which contains three essential ATP-binding sites. The function of the second motif, from residue 745 to 1286, remains confused. Thus, the *OPLAH* p.S323R would be located in the first domain extremely close to one of the ATP-binding sites (Fig. 1d), suggesting that this change in the human 5-oxoprolinase is probably affecting the proper activity of the enzyme. Regarding the *OPLAH* p.V1089I, this change would be placed in the second domain whose function is still unknown. This change has been recently annotated as SNP rs185836803. Other SNPs have been annotated in the public databases associated with diseases. Thus, the overwhelming majority of cases of familial amyloid polyneuropathy (FAP; MIM 105210) result from the p.V30M substitution in the *TTR* gene, which is referred to as SNP rs28933979 (Benson 2001). The fact is that the *OPLAH* p.V1089I is the only detected mutation in our patient and he suffers from a persistent oxoprolinuria.

Unfortunately, we could not measure oxoprolinase activity since patients did not collaborate with this study and at present are not controlled in the hospital. However, the homodimeric structure of the enzyme could make possible a dominant negative effect of the mutations, although these are rare in metabolic diseases. However, some forms of hypophosphatasia (MIM 146300, 241500, 241510) and also of cortisone reductase deficiency (MIM 604931) can be transmitted in a dominant manner and in these cases the mutations can exhibit a dominant-negative effect by inhibiting the enzymatic activity of the heterodimer (Lawson et al. 2011; Lia-Baldini et al. 2001).

These dominant-negative mutations are in the active site or in an area which probably affects the formation of functional dimers. A similar process could occur in heterozygous patients of 5-oxoprolinase deficiency. In some particular conditions, interactions between both identical molecules would not be properly made and would trigger a massive excretion of 5-oxoproline in urine. This anomalous biochemical condition would be normalized when the stressing conditions disappear, as in our proband ID no. 943. In other cases, the oxoprolinuria could persist, as in our proband ID no. 1037. Further studies are needed to clarify the possible mechanism causing this biochemical alteration in heterozygous individuals for *OPLAH* mutations. Independently, mutational analysis of the *OPLAH* gene may be advisable to elucidate causes of unknown massive pyroglutamic aciduria in order to understand better this biochemical alteration that may confound the final diagnosis.

Acknowledgments We thank all patients and their relatives for their kind collaboration.

A Concise One-Sentence Take-Home Message

Oxoprolinuria, due to heterozygous mutations in the *5-oxoprolinase* gene, may present as epiphenomenon in several pathological conditions and confound the final diagnosis.

Funding Support

C.E. has a "Miguel Servet" contract funded by the Instituto de Salud Carlos III (ISCIII). Centro de Investigación Biomédica en Red de Enfermedades Raras (CIBERER) is an initiative from the ISCIII.

Competing Interest

None declared.

References

Al-Jishi E, Meyer BF, Rashed MS, Al-Essa M, Al-Hamed MH, Sakati N, Sanjad S, Ozand PT, Kambouris M (1999) Clinical, biochemical, and molecular characterization of patients with glutathione synthetase deficiency. Clin Genet 55:444–449

Almaghlouth I, Mohamed J, Al-Amoudi M, Al-Ahaidib L, Al-Odaib A, Alkuraya F (2011) 5-Oxoprolinase deficiency: report of the first human OPLAH mutation. Clin Genet. doi:10.1111/j.1399-0004.2011.01728.x

Altschul SF, Gish W, Miller W, Myers EW, Lipman DJ (1990) Basic local alignment search tool. J Mol Biol 215:403–410

Benson MD (2001) The metabolic and molecular bases of inherited disease. In: Scriver CR, Beaudet AL, Sly WS, Valle D (eds) Amyloidosis. McGraw-Hill, New York, pp 5345–5378

Bernier FP, Snyder FF, McLeod DR (1996) Deficiency of 5-oxoprolinase in an 8-year-old with developmental delay. J Inherit Metab Dis 19:367–368

Cohen LH, Vamos E, Heinrichs C, Toppet M, Courtens W, Kumps A, Mardens Y, Carlsson B, Grillner L, Larsson A (1997) Growth failure, encephalopathy, and endocrine dysfunctions in two siblings, one with 5-oxoprolinase deficiency. Eur J Pediatr 156:935–938

Cooper DN, Chen JM, Ball EV, Howells K, Mort M, Phillips AD, Chuzhanova N, Krawczak M, Kehrer-Sawatzki H, Stenson PD (2010) Genes, mutations, and human inherited disease at the dawn of the age of personalized genomics. Hum Mutat 31:631–655

Dahl N, Pigg M, Ristoff E, Gali R, Carlsson B, Mannervik B, Larsson A, Board P (1997) Missense mutations in the human glutathione synthetase gene result in severe metabolic acidosis, 5-oxoprolinuria, hemolytic anemia and neurological dysfunction. Hum Mol Genet 6:1147–1152

Henderson MJ, Larsson A, Carlsson B, Dear PR (1993) 5-Oxoprolinuria associated with 5-oxoprolinase deficiency; further evidence that this is a benign disorder. J Inherit Metab Dis 16:1051–1052

Kumar A, Bachhawat AK (2010) OXP1/YKL215c encodes an ATP-dependent 5-oxoprolinase in Saccharomyces cerevisiae: functional characterization, domain structure and identification of actin-like ATP-binding motifs in eukaryotic 5-oxoprolinases. FEMS Yeast Res 10:394–401

Larsson A, Mattsson B, Wauters EA, van Gool JD, Duran M, Wadman SK (1981) 5-oxoprolinuria due to hereditary 5-oxoprolinase deficiency in two brothers–a new inborn error of the gamma-glutamyl cycle. Acta Paediatr Scand 70:301–308

Lawson AJ, Walker EA, Lavery GG, Bujalska IJ, Hughes B, Arlt W, Stewart PM, Ride JP (2011) Cortisone-reductase deficiency associated with heterozygous mutations in 11beta-hydroxysteroid dehydrogenase type 1. Proc Natl Acad Sci U S A 108: 4111–4116

Lia-Baldini AS, Muller F, Taillandier A, Gibrat JF, Mouchard M, Robin B, Simon-Bouy B, Serre JL, Aylsworth AS, Bieth E et al (2001) A molecular approach to dominance in hypophosphatasia. Hum Genet 109:99–108

Mayatepek E, Hoffmann GF, Larsson A, Becker K, Bremer HJ (1995) 5-Oxoprolinase deficiency associated with severe psychomotor developmental delay, failure to thrive, microcephaly and microcytic anaemia. J Inherit Metab Dis 18:83–84

Njalsson R (2005) Glutathione synthetase deficiency. Cell Mol Life Sci 62:1938–1945

Njalsson R, Carlsson K, Winkler A, Larsson A, Norgren S (2003) Diagnostics in patients with glutathione synthetase deficiency but without mutations in the exons of the GSS gene. Hum Mutat 22:497

Njalsson R, Ristoff E, Carlsson K, Winkler A, Larsson A, Norgren S (2005) Genotype, enzyme activity, glutathione level, and clinical phenotype in patients with glutathione synthetase deficiency. Hum Genet 116:384–389

Ristoff E, Larsson A (2007) Inborn errors in the metabolism of glutathione. Orphanet J Rare Dis 2:16

Roesel RA, Hommes FA, Samper L (1981) Pyroglutamic aciduria (5-oxoprolinuria) without glutathione synthetase deficiency and with decreased pyroglutamate hydrolase activity. J Inherit Metab Dis 4:89–90

Ruijter GJ, Mourad-Baars PE, Ristoff E, Onkenhout W, Poorthuis BJ (2006) Persistent 5-oxoprolinuria with normal glutathione synthase and 5-oxoprolinase activities. J Inherit Metab Dis 29:587

Schwahn B, Kameda G, Wessalowski R, Mayatepek E (2005) Severe hyperhomocysteinaemia and 5-oxoprolinuria secondary to antiproliferative and antimicrobial drug treatment. J Inherit Metab Dis 28:99–102

Shi ZZ, Habib GM, Rhead WJ, Gahl WA, He X, Sazer S, Lieberman MW (1996) Mutations in the glutathione synthetase gene cause 5-oxoprolinuria. Nat Genet 14:361–365

Printed by Publishers' Graphics LLC
MO20120920-274-243